Renewable Energy in Turkey
and Selected European Countries

Walter Leal Filho/Kerstin Kuchta
Franziska Mannke/Konstantin Haker

Renewable Energy in Turkey and Selected European Countries

Potentials, Policies and Techniques

A Handbook

With the collaboration of
Georgi Chobankov, Marko Gehrmann, Orhan Yenigün,
Turgut Onay, Burak Demirel, Ronald Wennersten
and Anna Spitsyna

PETER LANG
Internationaler Verlag der Wissenschaften

Bibliographic Information published by the Deutsche Nationalbibliothek
The Deutsche Nationalbibliothek lists this publication in the Deutsche
Nationalbibliografie; detailed bibliographic data is available in the internet at
<http://www.d-nb.de>.

Copy editing and lay-out:
Kumpernatz + Bromann
www.kumpernatz-bromann.de

ISBN 978-3-631-59922-8

© Peter Lang GmbH
Internationaler Verlag der Wissenschaften
Frankfurt am Main 2009
All rights reserved.

All parts of this publication are protected by copyright. Any
utilisation outside the strict limits of the copyright law, without
the permission of the publisher, is forbidden and liable to
prosecution. This applies in particular to reproductions,
translations, microfilming, and storage and processing in
electronic retrieval systems.

www.peterlang.de

Table of Contents

Abbreviations 7

Preface 11

Chapter 1
Potential for Renewable Energy in Turkey
Ronald Wennersten, Anna Spitsyna 13

Chapter 2
Renewable Energy Policy in Turkey, Germany and Sweden
Ronald Wennersten, Anna Spitsyna 29

Chapter 3
Best Available Techniques – Working Paper – Biogas
Kerstin Kuchta, Konstantin Haker 63

Chapter 4
Best Available Techniques – Working Paper – Biofuels
Kerstin Kuchta, Konstantin Haker, Georgi Chobankov 83

Chapter 5
Best Available Techniques – Working Paper – Photovoltaic
Kerstin Kuchta, Marko Gehrmann, Konstantin Haker 113

Chapter 6
Best Available Techniques – Working Paper – Waste to Energy (W2E) – Waste incineration –
Kerstin Kuchta, Konstantin Haker 147

Chapter 7
Best Available Techniques – Working Paper – Waste Management
Kerstin Kuchta, Konstantin Haker 179

Annex
Expertise Catalogue
Kerstin Kuchta, Konstantin Haker, Georgi Chobankov 209

About the Authors 229

Thematic Index 233

The content of this handbook is the sole responsibility of Bogaziçi University and can in no way be taken to reflect the views of the EU.

Abbreviations

$	US Dollar
€	Euro
AC/DC converter	Conversion of alternating current to direct current
AD	Anaerobic Digestion
APK	Research, Planning and Coordination Board of MENR
APM	Automatic Pricing Mechanism
ARE	Agency for Renewable Energy
a-Si	Amorphous Silicon
BAT	Best Available Techniques
BEWAG	Burgenländische Elektrizitätswirtschafts-Aktiengesellschaft
BMI	Business Monitor International
BO	Build-Operate
BOO	Build-Own-Operate
BOT	Build-Operate-Transfer
BREFs	Best Available Techniques REFerence Documents
CdTe	Cadmium-Telluride
CEAS	Cukurova Elektrik AS
CH_4	methane
CHP	Combined Heat and Power
CIS	Copper-Indium-di-Selenide
CO_2	Carbon Dioxide
COP	Coefficient of Performance
c-Si	Crystalline Silicon
ct	cent
Cz	Czochralski
DME	DiMethyl Ether Vehicle
DPT	The State Planning Organisation
EC	European Commission
ECU	European Currency Unit
EEG	Erneuerbare-Energien-Gesetz (Renewable Energy Sources Act)
EIE	The Turkish electrical power resources survey and development administration
EMRA	Energy Market Regulatory Agency

EMS	Environmental Management System
EPBD	Energy Performance of Buildings Directive
ETR	Ecological Tax Reform
EU	European Union
EVA	Ethylene-Vinyl-Acetate
FAL	Federal Agricultural Research Centre
FFV	Flexible Fuel Vehicle
FGT	Flue-Gas Treatment
G	Giga
GDEA	General Directorate of Energy Affairs
GDP	Gross Domestik Product
GHG	Greenhouse Gas
GW	Ground Wave
GWh	Gigawatt Hour
H_2S	hydrogen sulphide
HEW	Hamburg – Hamburger Elektizitätswerke
HIT	Heterojunction with Intrinsic Thin Layer
HTU	Hydrothermal Upgrading
IEA	International Energy Agency
IMF	International Monetary Fund
InGaAs	Indium-Gallium-Arsenide
InGaPh	Indium-Gallium-Phosphide
IPPC	Intergovernmental Panel on Climate Change
IWS	Ionisation Wet Scrubbers
KTH	Royal Institute of Technology
kWh	Kilowatt Hour
M	Million
m^3	Cubic Meter
MDGs	Millennium Development Goals
MENR	Ministry of the Environment and Natural Resources
MP	Member of Parliament
MPP Tracker	Maximum Power Point Tracks
MSEK	Million Swedish Crowns
MSW	Municipal Solid Waste
MTA	Turkish Mineral Research and Exploration Institute
Mtoe	Million Tons of Oil Equivalent

MW	Megawatt
MWe	Megawatts Electric
MWth	Thermal Megawatts
NAPs	National Action Plans
NE	Net Energy
NECC	National Energy Conservation Centre
NG	Natural Gas
NH_3	ammoniac
NOX	Nitrogen Oxides
OECD	Organisation for Economic Co-operation and Development
ppm	part per million
PPO	Pure Plant Oil
PV	Photovoltaic
R&D	Research and Development
RDF	Refuse Derived Fuel
REFIT	Renewable Energy Feed-in Tariff
RENET	Renewable Energy Networks between Turkish and European Universities
REO	Renewable Energy Obligations
RER	Renewable Energy Resources
RES	Renewable Energy Source
RESA	Reaktorschnellabschaltung
RES-E	Electricity Produced from Renewable Energy Sources
RF	Radio Frequency
RME	Rapeseed Methyl Ester
SCR	Selective Catalytic Reduction
SEE	State Economic Enterprises
SEK	Swedish Crowns (Currency)
SHP	Small Hydropower
SNCR	Selective Non-Catalytic Reduction
SPD	Social Democratic Party of Germany
STEM	Swedish National Energy Administration
StrEG	A federal Electricity Feed-In Law
T	Tera
TAEK	Turkish Atomic Energy Authority
TCO	Transparent Conductive Oxide

TETAS	Turkish Electricity Trading Company
TGC	Tradable Green Certificates
toe	Tons of Oil Equivalent
TOOR	Transfer of Operating Rights
TPA	Third Party Access
TWh	Terrawatt Hour
VAT	Value Added Tax
VOC	Volatile Organic Compounds
Wh	Watthour
WHPBA	The Wind Power and Hydropower Plant Businessmen's Association
WI	Waste Incineration

Preface

Renewable energy is an emerging field and one which congregates both technical developments and the economical use of energy resources. Yet, despite the relevance of this topic, there is today a wide lack of awareness, capacity and training in technological developments and in respect of the implementation of projects in the field of renewable energy. This is acknowledged as being a major obstacle in attempts to develop renewable energy systems. Weak market integration and imprecise regulatory frameworks are also some of the barriers seen both in European Union accession countries and elsewhere.

Despite the fact that more co-operation between the EU and Turkey in the field of renewable energy is needed and despite the willingness of organizations in both regions to pursue this, the amount of joint works in the field of renewable energy seen on the ground is still quite limited. Yet, much can be gained if institutions of higher education can co-operate in this field, especially in terms of education and training as well as in respect of knowledge and technology transfer.

This publication represents one of concrete outcomes of the EU project "Renewable Energy Networks between Turkish and European Universities (RENET)" which has over a period of 18 months established and sustained a network of Turkish and European universities in order to cater for a "university/university" and "university/ industry" interface. In particular, RENET aimed at improving the dialogue, mutual understanding and co-operation between German, Swedish and Turkish institutions of higher education in the key field of renewable energy – with a focus on waste-to-energy, solar energy and bio-fuels, in order to promote information and technology transfer among partners and stakeholders, hence assisting with the implementation of sustainable (energy) policies in both regions. Led by Bogazici University, Istanbul/Turkey, the project consortium further comprises the Hamburg University of Applied Sciences in Hamburg, Germany and the Royal Institute of Technology in Stockholm, Sweden as well as Akdeniz University in Antalya, Turkey.

This book is structured as follows: the first part will provide an overview of the current legislative frameworks for renewable energy in Turkey, Sweden and Germany. This will be followed by thematic chapters on key project issues such as solar energy, waste-to-energy and waste management, biofuels and biogas. The appendix provides a vast range of thematically organized sources that can be accessed to obtain further insights on the respective themes covered in this publication.

Considering all the above elements, the project RENET has contributed not only to the Millennium Development Goals (MDGs) on ensuring environmental sustainability and developing global partnerships for development, but it is also in line with the strategic objectives of the Lisbon-Gothenburg Strategies of the European Union which foster the innovation capacity and the creation of sustainable jobs within Europe. We hope this publication will not only illustrate the means via which renewable energy may be promoted within higher education institutions, but also foster more cooperation between universities in the European Union and in Turkey.

Hamburg, Istanbul, Stockholm 2009

Walter Leal Filho, Kerstin Kuchta, Franziska Mannke, Konstantin Haker, Georgi Chobankov, Marko Gehrmann, Orhan Yenigün, Turgut Onay, Burak Demirel, Ronald Wennersten, Anna Spitsyna

Chapter 1

Potential for Renewable Energy in Turkey

Ronald Wennersten, Anna Spitsyna

1. Introduction

Due to the social and economic developments in the country, the demand for energy in Turkey is growing rapidly, especially for electricity, which grew by 6,6% during 1995-2004.[1]

Turkey is very dependent on imported energy resources, such as oil, gas and coal, which places a great burden on the economy. Turkey's renewable energy sources are plentiful and extensive, and represent the second-largest domestic energy source after coal. Primary renewable energy resources in Turkey include hydropower, biomass, wind, biogas, geothermal and solar. The share of renewables in total electricity generation is 29,63%, while that of natural gas is 45% (year 2006 figures). The projection for the period 2006-2020 is for annual growth of 8% in total electricity generation.[2] The additional generation capacity needed up to 2020 will require huge investments (Demirbas 2002).

Electricity supply infrastructure in Turkey, as in many developing countries, is being rapidly extended, as policymakers and investors around the world increasingly recognise the essential role of energy in improving living standards and sustaining economic growth. The energy sector in Turkey is mainly state-owned, and the Government is heavily involved in the management and corporate decisions of the State Economic Enterprises (SEEs) (Demirbas 2002).

The potential of Turkey as a photovoltaic (PV) market is very large, since the country abounds in solar radiation and contains large areas of available land for solar farms. At present, Turkey has no appropriate legal framework to enable the production of more PV energy and the sale of excess energy to the grid. Therefore, most PV applications are for stand-alone power systems. The Turkish Government

---

[1] Turkish Weekly, The investment Potential of the Turkish Energy Market, Fevzi Saffet Bora, 4 February 2007.
[2] www.sciencedirect.com/science?_ob=ArticleURL&_udi=B6V4S-4S4S5JW-1&_user= 10&_coverDate=11%2F30%2F2008&_alid=799492189&_rdoc=1&_fmt=high&_orig= search&_cdi=5766&_sort=d&_docanchor=&view=c&_ct=7&_acct=C000050221&_version =1&_urlVersion=0&_userid=10&md5=a8dee198ad50ff5fa57c74ffcb7178e1.

needs to adapt the legal structure to include PV grid-connected power systems and to fund part of their cost.[3]

After lengthy discussions, the Turkish Parliament has approved a renewable energy law which will provide feed-in tariffs for electricity from renewable energy sources. The new renewable energy law will support renewable power by guaranteeing e.g. the average wholesale purchase price of electricity from wind power (some 5 ct/kWh) for a period of seven years for electricity generated from renewable sources.

- Solar power 0,18 €/kWh for 10 years;
- biomass 0,14-010, geothermal 0.07-0.06 €/kWh;
- hydro 0,05 €/kWh for 10 years.

2. General review of Turkey's renewable energy status

Due to the social and economic development in the country, the demand for energy and particularly for electricity is growing rapidly. The main native energy resources are hydro, mainly in the eastern part of the country, and lignite. Turkey has no major oil and gas reserves. Almost all oil, natural gas (NG) and high quality coal are imported, making Turkey a net energy importing country, with more than half its energy requirements being supplied by imports. Oil has the biggest share of total primary energy consumption. Because of recent efforts regarding diversification of energy sources, the use of natural gas, which was recently introduced into Turkey, has been growing rapidly.

Turkey's geographical location has several advantages for extensive use of most of its renewable energy sources. As mentioned above, hydropower energy is Turkey's main potential source of renewable energy. The other major areas of renewable energy development in Turkey are solar thermal, wind, geothermal and photovoltaic energy (Kaygusuz 2002a).[4]

---

[3] http://www.iea-pvps.org/ar/ar07/07ar_Turkey.pdf.
[4] http://www.dsi.gov.tr/english/congress2007/chapter_2/57.pdf.

Figure 1: Energy production by source, 1973-2020.

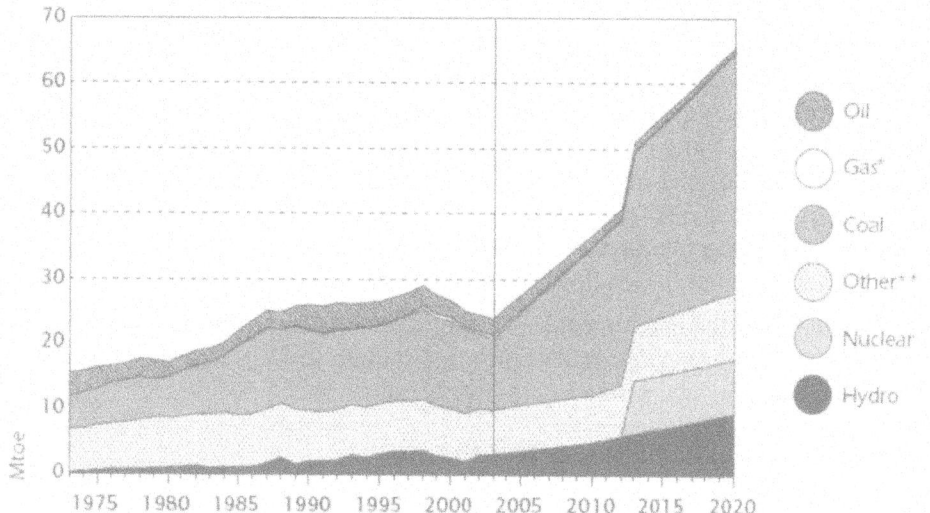

*Negligible.
**Includes geothermal, solar, wind, combustible renewables and waste.

Sources: Energy Balances of OECD Countries, IEA/OECD Paris, 2004; and country submission

The energy demand of Turkey is predicted to continue increasing strongly up to 2025. This rapid increase in demand is due to the high economic development rate of the country. The estimated amount of investment for energy production facilities by the year 2010 is around 45 billion dollars. Transmission and distribution facilities will require an additional 10 billion dollar investment in the same period. The Government has taken measures to attract new local and foreign private sector investments and also to transfer the operational rights of existing units to the private sector for their renewal and efficient operation.

3. Turkey's renewable energy potential and its utilisation

Global warming attributed to e.g. the use of fossil fuel is increasing in severity and is part of a growing climate problem that poses a danger to the common future of mankind. Hence, increasing electricity generation from renewable energy sources (green energy) is becoming more and more important on a global scale. In this context, EU countries are required to meet 21% of their energy needs from renewable energy sources by the year 2010, as stated in Directive 2001/77/EC (27 March 2001) on Promotion of Electricity Produced from Renewable Energy Sources in the International Electricity Market. To this end, all hydropower plants (including those with small capacity, less than 10 MW) have been included in an incentive plan. Since Turkey is a candidate for EU membership, it is obliged to take account of EU energy plans.

Practically all kinds of renewable energy sources are available in Turkey. However, apart for lignite and hydraulic energy, these sources are not sufficient to meet the energy requirements of the country, more than half of which is currently being met by imported energy. Table 1 shows the potential for renewable energy in Turkey. Due to the solar belt in which Turkey is located, its technical solar energy potential of 6.105 TWh/year is very high in terms of electricity production, followed by wind energy potential with an estimated value of 290 TWh/year and hydropower potential of 216 TWh/year (Table 1).

The renewables together currently provide 13,2% of the primary energy, mainly in the form of combustible renewables and wastes (6,8%), hydropower (about 4,8%) and other renewable energy resources (about 1,6%).[5]

Table 1: Renewable energy potential of Turkey

Renewable energy source	Form of energy usage	Natural potential	Technical potential	Economic potential
Solar energy	Electrical energy (TWh/year)	977.000	6.105	305
Hydraulic energy	Heat (Mtoe/year)	80.0000	500	25
Wind energy	Electrical energy (TWh/year)	433	216	127.4
Direct terrestrial	Electrical energy (TWh/year)	400	110	50
Direct maritime	Electrical energy (TWh/year)	–	180	–
Sea wave energy	Electrical energy (TWh/year)	150	18	–
Geothermal energy	Electrical energy (TWh/year)	–	–	1.4
Biomass energy	Heat (MW$_{th}$)	31.500	7.500	2.843
	Fuel (classic) (Mtoe/year)	30	10	7
	Fuel (modern) (Mtoe/year)	90	40	25

Source: A. Ozdamar, K.T. Gursel, G. Orer and Y. Pekbey, Investigation of the potential of wind–waves as a renewable energy sources: by the example of Cesme, Turkey, Renew Sustain Energy Rev 8 (2004), pp. 581–592

According to predictions provided by Business Monitor International (BMI), Turkish real GDP growth will average 6.13% per annum between 2007 and 2012, compared with the 2008 value of 4.8%. Total population is also expected to rise, and GDP per capita and electricity consumption per capita are predicted to increase significantly. Turkey's power consumption is expected to increase from an estimated 149 TWh in 2007 to 183 TWh by the end of 2012, while exports should rise from an estimated 36,3 TWh in 2007 to 71,6 TWh in 2012, assuming 6,4% annual growth in generation capacity.

---

[5] IEA 2007.

3.1 Solar

Mean annual solar radiation in Turkey is 3,6 kWh/m^2 and day and the total annual radiation period is around 2.640 h^6, sufficient for solar thermal applications. It has been calculated that Turkey receives sunlight equivalent to roughly 11.000 times the amount of electricity generated in Turkey in 1996[7]. Despite this colossal potential, flat-plate solar collectors for domestic hot water production in coastal regions are the only real use of solar energy. Solar energy thermal potential is 61 Mtoe and electricity potential is 15 Mtoe.

The potential of Turkey as a photovoltaic market is very large, since the country abounds in solar radiation and has large areas of available land for solar farms. At present, Turkey has no appropriate legal framework to enable the production of more PV energy and the sale of excess energy to the grid. Therefore, the most PV applications are for stand-alone power systems. The Turkish Government needs to adapt the legal structure to include PV grid-connected power systems and to fund a part of their cost.[8]

3.2 Wind

In 2000, the Government of Turkey offered tenders for up to 390 MWe of electricity from windpower. About 25 potential sites for windpower projects had been identified and were undergoing evaluation, but the tender was cancelled as part of the IMF-induced economic policy changes. Without the full sovereign guarantees that would in effect result in government subsidies to offset the relatively high expected cost of the power produced, none of 17 windpower projects that had received their Build-operate-transfer (BOT) approvals have proceeded.

In December 2006, the Ministry of Energy published a wind map of Turkey, which has stimulated wind power investments. The Wind Power and Hydropower Plant Businessmen's Association (WHPBA) puts the potential Turkish wind power figure at 10,0 GW.[9]

According to a May 2008 report from the Global Wind Energy Council, Turkey is one of the best potential markets for wind in Europe because of the high wind speeds. The country had 146 MW of wind power at the end of 2007, with an ad-

---

[6] General Directorate of Electrical Power Resources Survey and Development Administration (EIE). Renewable energy resources, solar energy, solar energy studies, http://www.eie.gov.tr/turkce/gunes/eiegunes.html, 2006.

[7] http://turkey-electricity.com/page12.html.

[8] http://www.iea-pvps.org/ar/ar07/07ar_Turkey.pdf.

[9] Turkey Electricity-Renewable Energy, http://turkey-electricity.com/page9.html.

ditional 600 MW scheduled to come online by the end of 2009.[10] Turkey has a goal of deriving 2% of its electricity from wind power.

3.3 Geothermal energy

Turkey has significant potential for geothermal power production, and is considered the seventh-richest country in the world in this area. The first geothermal research and investigations in Turkey were started by the Turkish Mineral Research and Exploration Institute (MTA) in the 1960s. Much of this potential is of relatively low enthalpy that is not suitable for electricity production, but is still useful for direct heating applications. By the end of 1999, Turkey's total installed capacity for direct heating was 820 thermal MW, of which about 390 MW provided heating for 51.600 residences, about 100 MW provided heating for about 45 hectares of greenhouses, and about 330 MW were used to provide heated water for about 200 spas. By 2010, as many as 500.000 residences could be heated by geothermal power, which would represent the use of about 3.500 MW (FEI 2003).

The overall geothermal energy potential of Turkey is estimated at 35 GW per year. However in 2001, geothermal energy production was only 1.759 Mtoe. In December 2007 the Turkish Government enacted the Application Regulation of the Law on Geothermal Resources and Natural Mineral Water. This Regulation covers the process, elements and sanctions regarding:

- Issuing a licence operation;
- transferring this licence;
- auditing the actions;
- resources and the environment;
- revoking the licence;
- protecting the resources;
- leaving the area of the licence in relation to the geothermal resources and natural mineral water that are specified or will be specified and gases originating from geothermal sources.

Geothermal energy can be utilised in different forms such as electricity generation, direct use, space heating, greenhouse heating and industrial usage. Today in Turkey, biomass and hydropower are mainly used, with geothermal in third place. Geothermal electricity generation has a minor role in Turkey's electricity capacity, as low as 0,09%, but the projections foresee an improvement to 0,32% by the year 2020.

---

[10] http://www.windfair.net/press/5130.html.

Turkey currently has two operating geothermal power plants, a 15 MW facility in the Denizli-Kizildere geothermal field in the south-western province of Denizli that includes nine production wells and also has an integrated liquid carbon dioxide (CO_2) and dry ice production factory that can produce a combined total of 40.000 metric tons per year of the two products, and another facility with a capacity of 8 MW. Six other geothermal fields have been identified, all in the extreme south-west of Turkey, that may be suitable for geothermal power production: the Germencik-Aydin field in Aydin Province, the Çanakkale-Tuzla field in Çanakkale Province, the Izmir-Sefirihiser field in Izmir Province, the Aydin-Salvatli field in Aydin Province, the Kutahya-Simav field in Kutahya Province, and the Dikili-Bergama field in Izmir Province. The Germencik-Aydin field may be the most promising of these, as it has a power potential of at least 100 MWe.[11]

3.4 Hydropower

Hydropower is likely to become an important energy source in Turkey. Turkey's hydro-electric potential can meet up to 46% of its electric energy demand by 2020 and this potential can be developed simply and economically.[12]

There are 678 sites available for hydroelectric plant construction, 135 of which are already being developed, distributed over 26 main river zones. The total gross potential of these sites is nearly 37 GW and the total energy production capacity 127 TWh/yr, about 30% of which may be economically usable. At present only 35% of the total hydroelectric power potential (around 13.000 MW) is operational. The national development plan aims to increase this to 100% by 2010. The input of small hydroelectric plants to total electricity generation is estimated at 5-10%.

Turkey's gross theoretical hydroelectricity potential is 433 billion kWh, which is almost 1% of the theoretical global potential and 14% of the European potential.[13]

The overall hydropower potential in Turkey is 190 billion kWh per year, but this may decrease to 130 billion kWh due to climate change.[14] At present there are 150 hydropower plants with a combined installed capacity of 13,0 GW, a total capacity of 3,0 GW is under construction, and 20,6 GW is in the planning process.[15]

At the end of 2001, Turkey had 125 hydroelectric power plants in operation, ranging in size from the 2.400 MWe Atatürk Power Plant (currently the sixth

---

[11] http://www.turkey-electricity.com/page6.html.
[12] http://www.dsi.gov.tr/english/congress2007/chapter_2/57.pdf.
[13] http://www.dsi.gov.tr/english/congress2007/chapter_2/27.pdf.
[14] Turkish Weekly, The Investment Potential of the Turkish Energy Market, Fevzi Saffet Bora, 04 February 2007.
[15] Turkey Electricity-Renewable Energy, http://turkey-electricity.com/page9.html.

largest capacity hydroelectric facility in the world) all the way down to many small facilities of less than 2 MWe in capacity. Most are owned and operated by independent companies, including Birecik AS, which owns a 672 MWe power plant on the Euphrates River, and Cukurova Elektrik AS (CEAS), which currently has more than 1.000 MWe generating capacity. The Turkish Government hopes to see hydroelectric capacity expanded to 35.000 MWe by the year 2010. Ultimately, the construction of more than 300 additional hydroelectric power plants is projected for Turkey to make use of the remaining possible hydroelectric sites, which have a potential of about 69.000 GWh per year. This long-term plan would bring an additional 19.300 MWe of hydroelectric capacity online at a cost of more than $30 billion.[16]

3.5 Bioenergy

Biomass also represents a significant share of total energy consumption in Turkey, despite a drop from 20% in 1980 to 8% in 2005. Bioenergy represents about two-thirds of renewable energy production in Turkey. Various agricultural residues available in Turkey, such as grain dust, wheat straw and hazelnut shells, are the main source of biomass energy. Among the biomass energy sources, fuelwood seems to be one of the most interesting. The total forest potential of Turkey is around 935 million m^3, with an annual growth of about 28 million m^3.[17]

The total recoverable bioenergy potential was estimated at 6,98 Mtoe in 2000 and 7,26 Mtoe in 2005. These estimates were based on the recoverable energy potential from agricultural residues, livestock farming wastes, forestry and wood processing residues and municipal wastes. Total biomass production is expected to be 7,52 Mtoe in 2020 (Table 2).[18]

Table 2: Current and planned biomass energy production in Turkey

Year	Total biomass production (Mtoe)
2000	6,98
2005	7,26
2010	7,41
2020	7,52

Source: http://cat.inist.fr/?aModele=afficheN&cpsidt=20053566

---

[16] http://traccess.tubitak.gov.tr/fp6_yeni/DefaultIframe_en.aspx?aId=529.
[17] http://traccess.tubitak.gov.tr/fp6_yeni/DefaultIframe_en.aspx?aId=529.
[18] http://www.thetravelfoundation.org.uk/assets/tools_training_guidelines/policy%20papers/turkey.pdf.

Biogas potential has been estimated at 1,5-2 Mtoe. However, only two small units are in operation (5 MW) and one new facility (1 MW) has been licensed.

4. Links with political developments in the country

Government incentives play an important role in making renewable energy more attractive. The Turkish Government is in the process of passing a new energy law to require retail electricity licensees to purchase at least 8% of their annual electricity sales volume from renewable energy resources. The purpose of this law is to expand the use of renewable energy sources for electricity production purposes and to make them economically viable. The law covers renewable energy sources, i.e. other than fossil, such as hydraulic, wind, solar, geothermal, biomass, biogas, wave, flow energy and ebb and flood.[19]

There are two incentive schemes in general use world-wide: renewable energy feed-in tariffs (REFITs) and renewable energy obligations (REO). REFIT refer to favourable fixed or variable prices for renewable energy, while REO require utilities to supply a certain percentage of their electricity from renewable sources.

Turkey has adopted a hybrid system. Renewable power plants built before 2012 (the deadline set by law, which may be extended for another two-year period) are eligible for a REFIT of €50-55 per MWh (≈ USD 79 per MWh, compared with USD 82 regulated electricity sale price after tax and levies) for the first 10 years of operation, providing a hedge against foreign exchange risks. Furthermore, retail licence owners are required to allocate (as REO) a portion of their electricity purchases to renewable power.[20]

According to the Turkish Renewable Energy Act, all non-fossil based energy sources are considered a renewable energy source. Most preferred generation types, such as wind power, river-run hydropower plants and dammed hydro plants with reservoir areas smaller than 15 km^2 have been identified as renewable energy sources.[21]

One of the most important developments in the Turkish energy sector was the implementation of build-operate-transfer (BOT) and BO (build-operate) model investments with foreign capital. In 2001 these financing schemes were replaced by financial incentives within the framework of the Electricity Market Law.

---

[19] http://www.turkey-electricity.com/page15.html.
[20] http://www.turkey-electricity.com/page9.html.
[21] http://emeraldinsight.com/Insight/ViewContentServlet;jsessionid=59946E1A6A837378 677E0CC29B467EC3?Filename=Published/EmeraldFullTextArticle/Articles/0830170502.html.

According to the Electricity Market Licensing Regulation, promotion of RES has been assigned to the Energy Market Licensing Regulation (EMRA). This regulation aims to use RES with a reduced licence fee of 1% for the construction of facilities based on natural resources and renewable energy resources.

The Energy Market Regulatory Agency (EMRA) grants licence fee exemptions for renewable energy investors and the Turkish Electricity Trading Company (TETAS) can provide purchase guarantees to renewable energy companies. Renewable energy will play an important role as Turkey's preparations for accession to the European Union get underway.

Wind power is also attracting the attention of many companies. EMRA has issued 59 wind farm licences, with a total capacity of 856 MW. However, only 32 MW of that capacity is currently operational. In December 2006, the Ministry of Energy published the wind map of Turkey, which has generated investment in wind power. The Wind Power and Hydropower Plants Businessmen's Association (WHPBA) puts the potential Turkish wind power figure at 10,0 GW. The move away from natural gas caused by high prices has benefited renewable energy the most, particularly hydro and wind generation.

Current Turkish law and regulations on using renewable energy sources comprise two pieces of legislation: the Electricity Market Licensing Regulation and the Law on Utilisation of Renewable Energy Sources. These pieces of legislations were developed for the electricity sector.

Turkish energy policy is mainly concentrated on ensuring the supply of energy reliably, sufficiently, in time, in economic and clean terms and in a way to support and orientate the target growth and social developments. Within this framework, energy planning studies for the country, taking account of short-, medium- and long-term policies and measures, have been carried out by the Ministry of the Environment and Natural Resources (MENR). The Government has focused its efforts on improvement of domestic production by utilising public, private and foreign sources and increasing efficiency by rehabilitation and acceleration of existing construction programmes to initiate new investments on reasonable accounts.

Liberalisation of the energy sector and privatisation activities has been carried out. In order to overcome financial constraints, some incentives were provided through formulas such as BOT BO and TOOR (Transfer of Operating Rights) in the electricity sector.

The main principles of Turkish energy policy are:

- Meeting demand, making use of domestic sources in the first place as much as possible in all energy types;
- developing existing sources while accelerating studies on new sources;

- adding new and renewable sources as soon as possible to the energy supply cycle;
- source diversification and avoiding dependence on a single source or country in energy importation;
- meeting demand in the medium and long term through public, private and foreign capital contribution;
- encouraging private sector investment and expediting privatisation activities in the power sector;
- in the selection of new production projects, determining priorities that would respond to the national requirements in the shortest time and the most economical way and, besides hydraulic projects, taking into consideration thermal plants using different types of fuels and making the necessary preparations to introduce nuclear technology;
- upgrading existing transmission and distribution lines to supply reliable electric power of high quality;
- implementing energy-saving practices for energy efficiency, preventing waste and minimising losses in energy production, transmission and consumption;
- protecting the environment and public health in the process of meeting the energy requirements;
- continuing to encourage research and development in energy efficiency, renewable energy sources and especially in the area of applied research.

On the other hand, in order to establish and implement an integrated environmental energy production and consumption policy, a dynamic energy-environment policy is being structured to guard national interests while at the same time fulfilling international commitments.

Taking into account its geopolitical location, Turkey has the potential to become a very important terminus point for oil and gas exports and the priority goal of the Government is to become the 'Eurasian Energy Corridor' of the 21st Century.

Energy security is another important priority in Turkish Government policy in order to meet the domestic energy demand safely. In addition to securing resource security by establishing storage and stock facilities, all necessary precautions for energy and resource diversification are considered essential. Improvement studies on resource diversification have been carried out in cooperation with neighbouring countries.[22]

---

[22] http://traccess.tubitak.gov.tr/fp6_yeni/DefaultIframe_en.aspx?aId=529.

5. Difficulties, possible solutions, essential reforms

5.1 Barriers

The main barriers to developing renewable energy are:

- Lack of financial resources and appropriate lending facilities, particularly for small-scale projects;
- lack of detailed renewable energy resource assessments and databanks pertaining to Turkey (as to many other countries).

However, lack of awareness and knowledge is not a huge barrier in Turkey. Renewable energy is recognised as a major potential for indigenous, clean energy production. The most important handicap for foreign investors is Turkish bureaucracy. Permit applications by foreign investors can take up to a year, with numerous authorities being involved. The new Government had promised to simplify this permit application process.

5.2 Possible solutions

Hydroelectric generation, biomass combustion, solar energy for agricultural grain drying and hot water heating, and geothermal energy have been in use in the country for many years. Domestic water heating is the primary active solar technology. In Turkey, approximately 30.000 solar water heating systems have been installed since the 1980s. This is a minute fraction of the total potential, as about 50% of existing dwellings could be fitted effectively with a solar water heater. If this potential were extended to 2025, the deployment of approximately 5 million systems (allowing for a rise in the Turkish housing stock) would be required. This could save an estimated 30 PJ (9,0 TWh) per year of oil, coal and gas and 2,0 TWh per year of electricity, giving a saving of 5,0 million tonnes of CO_2 per year, or just under 1% of current Turkish CO_2 production.

Agricultural residues have a high potential to take the place of the lignite (40 million tons) and hard coal (1,3 million tons) used in electricity production.

Biogas systems are considered to be strong alternatives to traditional space heating systems (stoves) in rural Turkey. Geothermal heat pumps are a relatively new application of geothermal energy that has grown rapidly in recent years. The greatest benefit of geothermal heat pumps is that they use 25-50% less electricity than conventional heating or cooling systems. Geothermal heat pumps can also reduce energy consumption and corresponding air pollution emissions, up to 44% compared with air source heat pumps and up to 72% compared with electric resistance heating with standard air conditioning equipment.

5.3 Essential reforms

Turkey lacks a clear strategy concerning its renewable energy sources because of energy costs and investment costs. The Government encouraged the private sector to invest in natural gas combined-use circuit plants and guaranteed to buy the electricity generated at a low cost and with special conditions.

Turkey is interested in renewable energy resources and is devoting efforts to ensure the sustainability of using these energy resources. The Government encouraged the municipal authorities in respect of geothermal energy and gave them self-governing powers in this regard.

In Turkey, the efficiency of energy utilisation is not yet as high as it is in Europe. The Government is asking the private sector to supplement World Bank credit as regards all sources of renewable energy. The Government has agreed to act as guarantor for 30-40% of the cost of private sector investments to meet their own energy needs. If the private sector can find a buyer, it may sell the excess electricity produced in these plants.

It is only recently that less energy-consuming building projects have begun to be introduced and ground-source heating and passive heating systems are still uncommon.

6. Conclusions

Turkey has a substantial renewable energy potential and is keen to reduce its dependence on fossil fuels by increasing its use of Renewable Energy Resources (RER).

The wind sector is a good example of the increasing interest in generation of electricity with renewable resources. By the end of 2007, the EMRA has issued 751 license applications for RER electricity generation projects, with a total capacity of approximately 78.000 MW. It is expected that the Government will call for tenders for wind licences. A highly competitive market is emerging in Turkey and there will be further opportunities for foreign investors to enter it, both as direct investors and as partners with local companies that have already obtained generation licences.[23]

Achieving even the modest environmental goals of the Kyoto Protocol requires the sustained and orderly commercial development of viable renewable energy options. It is not enough for governments to support the development of renewable energy technologies. They must also support their commercial application in the country. In the case of Turkey, renewable energy resources do not yet have wide

---

[23] http://www.mondaq.com/article.asp?articleid=59066.

applications due to some technological and economic constraints. However, renewable energy usage by government and private companies is likely to increase year by year because Turkey is an energy-importing country, domestic fossil fuel resources are limited and the economic conditions of the country are not ideal. This report shows that there is huge potential for renewable energy in Turkey, especially hydropower, biomass, geothermal, solar and wind.

In conclusion, a step-wise (i.e. combined-use) shift from fossil fuels to renewable energy sources seems to be the obvious alternative for Turkey.[24]

7. References

General Directorate of Electrical Power Resources Survey and Development Administration (EIE). Renewable energy resources, solar energy, solar energy studies, http://www.eie.gov.tr/turkce/gunes/eiegunes.html, 2006.

http://cat.inist.fr/?aModele=afficheN&cpsidt=20053566.

http://emeraldinsight.com/Insight/ViewContentServlet;jsessionid=59946E1A6A 837378677E0CC29B467EC3?Filename=Published/EmeraldFullTextArticle/ Articles/083017050 2.html.

http://traccess.tubitak.gov.tr/fp6_yeni/DefaultIframe_en.aspx?aId=529.

http://turkey-electricity.com/page12.html.

http://www.dsi.gov.tr/english/congress2007/chapter_2/27.pdf.

http://www.dsi.gov.tr/english/congress2007/chapter_2/57.pdf.

http://www.iea-pvps.org/ar/ar07/07ar_Turkey.pdf.

http://www.mondaq.com/article.asp?articleid=59066.

http://www.sciencedirect.com/science?_ob=MiamiImageURL&_imagekey= B6V2P-458 N8PC-2-8&_cdi= 5708&_user= 4478132&_check= y&_orig= search&_coverDate= 02%2F 28%2F2003&view= c&wchp= dGLbVzz-zSkWb&_valck= 1&md5=89757551f86e8fda48 baecfb1f05fccf&ie= /sdarticle.pdf.

http://www.thetravelfoundation.org.uk/assets/tools_training_guidelines/policy% 20papers/ turkey.pdf.

http://www.turkey-electricity.com/page6.html.

http://www.turkey-electricity.com/page9.html.

http://www.turkey-electricity.com/page15.html.

---

[24] http://www.sciencedirect.com/science?_ob=MiamiImageURL&_imagekey=B6V2P-458 N8PC-2-8&_cdi=5708&_user=4478132&_check=y&_orig=search&_coverDate=02%2F 28%2F2003&view=c&wchp=dGLbVzz-zSkWb&_valck=1&md5=89757551f86e8fda48 baecfb1f05fccf&ie=/sdarticle.pdf.

http://www.windfair.net/press/5130.html.

International Energy Agency (IEA) (2007).

Ozdamar, A., Gursel, K.T., Orer, G. and Pekbey, Y. (2004). Investigation of the potential of wind–waves as a renewable energy sources: by the example of Cesme, Turkey, Renew Sustain Energy Rev 8 (2004), pp. 581–592.

Turkey Electricity-Renewable Energy, http://turkey-electricity.com/page9.html.

Turkish Weekly, The Investment Potential of the Turkish Energy Market, Fevzi Saffet Bora (04 February 2007).

www.sciencedirect.com/science?_ob= ArticleURL&_udi= B6V4S-4S4S5JW-1&_user= 10&_coverDate= 11%2F30%2F2008&_alid= 799492189&_rdoc= 1&_fmt= high&_orig= search&_cdi= 5766&_sort=d&_docanchor= &view= c&_ct= 7&_acct= C000050221&_version = 1&_urlVersion= 0&_userid= 10&md5= a8dee198ad50ff5fa57c74ffcb7178e1.

Chapter 2

Renewable Energy Policy in Turkey, Germany and Sweden

Ronald Wennersten, Anna Spitsyna

1. Introduction

In January 2007, the European Commission adopted an energy policy for Europe. This was supported by several documents on different aspects of energy and included an action plan to meet the major energy challenges Europe. Since Turkey is a candidate for EU membership, it is obliged to take account of such EU energy plans.

EU policies have imposed disproportionate burdens on Member States. For example, the 1997 White Paper Energy for the Future: Renewable Sources of Energy states that 'the overall EU target of doubling the share of renewables to 12% by 2010 implies that the Member States have to encourage the increase of RES [renewable energy sources] according to their own potential'.

More recently, legislation passed by the European Parliament has obligated Member States to adopt national targets for expansion of the renewable energy share. The 2001 Directive on the Promotion of Electricity Produced from Renewable Energy Sources in the Internal Electricity Market (RES-E) aims to increase the share of renewable energy to 12% of primary energy consumption in the EU and to boost renewable sources to 22% of electrical power production by 2010. The Directive mandates that 'Member States shall take appropriate steps to encourage greater consumption of electricity produced from renewable sources in conformity with the national indicative targets (Table 1).

The European Commission is promoting the deployment of renewable energy technologies across the EU to address three main energy policy challenges. First, under the Kyoto Protocol, it has agreed on EU-wide reductions in greenhouse gas emissions of 15% from 1990 levels in the first reporting period, 2008-2012.

Table 1: Reference Values for EU Member States' National Targets for the Contribution of Renewable Energy Sources to Gross Electricity Consumption by 2010

	TWh 1997	% Electricity Fuel Mix 1997	% Electricity Fuel Mix 2010 target
Belgium	0,86	1,1	6,0
Denmark	3,21	8,7	29,0
Germany	24,91	4,5	12,5
Greece	3,94	8,6	20,1
Spain	37,15	19,9	29,4
France	66,00	15,0	21,0
Ireland	0,84	3,6	13,2
Italy	46,46	16,0	25,0
Luxembourg	0,14	2,1	5,7
Netherlands	3,45	3,5	9,0
Austria	39,05	70,0	78,1
Portugal	14,30	38,5	39,0
Finland	19,03	24,7	31,5
Sweden	72,03	49,1	60,0
United Kingdom	7,04	1,7	10,0

Source: Directive 2001/77/EC of the European Parliament and of the Council of 27 September 2001 on the promotion of electricity produced from renewable energy sources in the internal electricity market, OJ L 283 27.10.2001, 39

A second set of energy considerations surrounds the high, and steadily rising, level of European energy import dependence. Imports account for 50% of EU energy consumption today and are expected to rise to 70% by 2020 in the absence of policies to curtail them. In Turkey this is an even greater problem (80% dependence), although Turkey possesses its own resources. The EU Commission has also noted that the development of renewable energy can be a valuable instrument for economic development and greater social cohesion within the EU. Since renewable energy technologies are developed and manufactured in several EU countries, renewable energy growth, in the Commission's estimation, can foster job creation, technical capacity development, and international research collaboration across Europe.

This report provides a detailed summary of the current situation for renewable energies in the RENET partner countries Turkey, Germany and Sweden (Section 2). The renewable energy policy outlook (Section 3) then identifies existing gaps and highlights future development of the European renewable energy policy and the national renewable policies of the respective countries in particular.

2. Renewable Energy Country Profile

2.1 Turkey

The real beginning for renewable energy policy in Turkey was the definition of renewable energy sources in the decree of the Modification of the Licence Regulation in the Electricity Market in 2003. Before then, there was no national renewable energy policy and few government incentives existed to promote market deployment of renewable energy. However, the Electricity Market Licensing Regulation, in itself, is not expected to be sufficient to overcome the high investment cost, risk and lack of security associated with the entry of renewable power plants into the electricity market.

The industrial sector in Turkey accounted for 40% of total final energy consumption and for 54% of electricity consumption in 2000, while the agriculture, household and services sectors together accounted for 40% of final energy consumption and 46% of electricity consumption. Although all four sectors have important potential for energy conservation, industry has been targeted as a priority area for energy conservation programmes owing to the projected rapid expansion of industrial energy demand. On the other hand, the structure of industry in Turkey is energy-intensive.[1]

Turkey's renewable energy sources are plentiful and extensive and represent the second-largest domestic energy source after coal. Primary renewable energy resources in Turkey are: hydro, biomass, wind, biogas, geothermal and solar.

The renewable energy supply in Turkey is dominated by hydropower and biomass, but environmental and scarcity-of-supply concerns have led to a decline in biomass use, mainly for residential heating. The total renewable energy supply declined from 1990 to 2004, due to a decrease in biomass supply. As a result, the composition of renewable energy supply has changed and wind power is beginning to claim market share. As a contributor to air pollution and deforestation, the share of biomass in the renewable energy sector is expected to decrease with the expansion of other renewable energy sources.[2]

2.1.1 Hydropower

There are 436 sites available for hydroelectric plant construction in Turkey, distributed on 26 main river zones. The total gross potential of these sites is nearly 50 GW and the total energy production capacity 112 TWh/yr, and about 30% of the

---

[1] http://www.planbleu.org/publications/atelier_energie/TR_Summary.pdf.
[2] Global warming and renewable energy sources for sustainable development: A case study in Turkey: Selçuk Bilgen, Sedat Keleş, Abdullah Kaygusuz, Ahmet Sarı and Kamil Kaygusuz 2006.

total gross potential may be economically exploitable. At present, only about 35% of the total hydroelectric power potential is in operation. The national development plan aims to harvest all of the hydroelectric potential by 2010. The contribution of small hydroelectric plants to total electricity generation is estimated to be 5-10%.[3]

2.1.2 Bioenergy

Among the renewable energy sources, biomass is important because its share of total energy consumption is still high in Turkey. However, the contribution of biomass resources to total energy consumption dropped from 20% in 1980 to 8% in 2005. Biomass in the form of fuelwood and animal wastes is the main fuel for heating and cooking in many urban and rural areas.

Using vegetable oils as an alternative fuel has economic, environmental, and energy benefits for Turkey. Vegetable oils have heat contents that are approximately 90% of that of diesel fuel. A major obstacle deterring their use in the direct-injection engine is their inherent high viscosity, which is nearly ten times that of diesel fuel. An overall evaluation of the data indicates that these oils and bio-diesel can be proposed as possible candidates for fuel. On the other hand, organic wastes are of vital importance for maintaining soil fertility, but in Turkey most of these organic wastes are used as fuel through direct combustion.

Biogas systems are considered to be strong alternatives to the traditional space heating systems (stoves) in rural Turkey. A recent study compared the economics of biogas systems with those of traditional heating systems fuelled by wood, coal/wood mixtures and dried animal waste in three different climatic regions of the country.The technical data used in the analysis were based on experimental results, and payback periods, cumulated life-cycle savings and the cost of biogas were calculated for a wide range using two variable economic parameters, discount and inflation rates. Despite reservations regarding the quality of the model and the assumptions, the results provide evidence of the economic viability of biogas systems over the traditional space heating systems of rural Turkey in many instances.

2.1.3 Geothermal energy

Turkey is one of the countries with significant potential in geothermal energy. Data accumulated since 1962 show that there may be about 4500 MW of geothermal energy usable for electrical power generation in high enthalpy zones. Heating capacity in the country runs at 350 MWt equivalent to 50.000 households. The total geothermal energy potential of Turkey is about 2.268 MW, but

---

[3] Global warming and renewable energy sources for sustainable development: A case study in Turkey: Selçuk Bilgen, Sedat Keleş, Abdullah Kaygusuz, Ahmet Sarı and Kamil Kaygusuz 2006.

the share of geothermal energy production, both for electrical and thermal uses is only 1.229 MW.

2.1.4 Solar energy

Turkey lies in a sunny belt between 36,1 and 42,1 °N latitudes. The mean annual solar radiation is 3,6 kWh/m^2 and day and the total annual radiation period is approximately 2.640 h, which is sufficient to provide adequate energy for solar thermal applications. In spite of this high potential, solar energy is not yet widely used, except for flat-plate solar collectors. They are only used for domestic hot water production, mostly in the sunny coastal regions.

2.1.5 Wind energy

There are a number of cities in Turkey with relatively high wind speeds. These have been classified into six wind regions, with a low of about 3,5 m/s and a high of 5 m/s at 10 m altitude, corresponding to a theoretical power production between 1.000 and 3.000 kWh/m^2 and yr. The most attractive sites are the Marmara Sea region, Mediterranean Coast, Aegean Sea Coast and Anatolian inland. Turkey's first wind farm was commissioned in 1998 and has a capacity of 1,5 MW.

Figure 1: Total Primary Energy Supply From Renewables in Turkey.

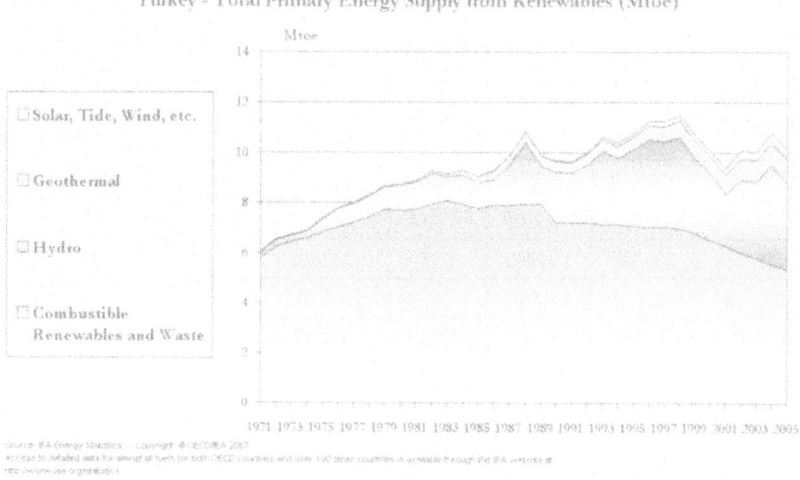

Source: IEA Energy Statistics – Copyright © OECD/IEA 2007; www.iea.org/statistics

The Turkish electrical power resources survey and development administration (EIE) is carrying out wind measurements at various locations to evaluate wind energy potential over the country, and has started to compile a wind energy atlas.

2.2 Germany

Between 1990 and 2003, the renewable energy share in Germany's electrical power generation fuel mix grew from less than 3% to almost 9%. Over the same period, net electricity consumption in Germany grew by approximately 5%, while carbon dioxide emissions from electric power production declined by roughly 13%. Germany produces nearly half of all the wind turbines in the world, and a third of all solar (photovoltaic) cells. The German Government sees renewable energy as an inevitable trend and wants to put Germany at the top of the international market.[4]

Since 1997, Germany and the other Member States of the European Union have been working towards a target of 12% renewable electricity by 2010. This target was already surpassed in Germany in 2007 when the renewable energy share reached 14%. On April 26, 2007, Environment Minister Sigmar Gabriel announced that this target would rise to 27% by 2020. Electricity use is to be cut by 11%, and the number of co-generation plants is to double.

Table 2: Increase in electricity production from renewable energy in Germany 1990-2007 in GWh

Year	Hydro-power	Wind-power	Biomass	Landfill biogas	Photo-voltaic	Geothermal power	Total	Share
1990	17.000	40	222	1.200	0.001	0	18.463	3,4%
1995	21.600	1.800	670	1.350	0.011	0	25.431	4,7%
2000	24.936	7.550	2.279	1.850	0.064	0	36.679	6,3%
2001	23.383	10.509	3.206	1.859	0.116	0	39.073	6,7%
2002	23.824	15.786	4.017	1.945	0.188	0	45.760	7,8%
2003	20.350	18.859	6.970	2.162	0.313	0	48.654	7,9%
2004	21.000	25.509	8.347	2.116	0.557	0,2	57.529	9,3%
2005	21.524	27.229	10.495	3.039	1.282	0,2	63.569	10,4%
2006	20.000	30.700	15.490	3.639	2.220	0,4	72.049	11,7%
2007	20.700 21.000	39.500 38.500	19.500 + biogas	4.250 18.500	3.500 3	0,4 0,1	87.450	14,2%

Source: BMU based on AGEE -Stat and ZSW [3]; EnBW [12]; BWE [16]; StBA [5]; BMELV [75]; IE [8], [13], [20], [70]; VDN [9]; BDEW [71]; AGEB [2], [18]; FN R [7]; SF V [28]; BSW [10]; ZfS [19]; Erdwärme-Kraft [79]; DEWI [62], [67], [76], [77], [78]; geox GmbH [66]

Onshore wind generation in Germany totalled 25.509 GWh in 2004. Together with hydro power (21.076 GWh), it dominates the market. Biomass electricity

---
[4] http://www.wealthdaily.com/articles/renewable-energy-germany/1418.

also made a significant contribution (9.326 GWh), with strong growth in the solid biomass sector (average annual growth of 34% between 1997 and 2004). Onshore wind did a little bit better (36%), but was left behind by PV, which increased on average by 53% per year (Table 2).

In absolute figures, however, the largest growth occurred in the onshore wind power sector (56.468 GWh in 2004 compared with 23.843 GWh in 1997).

2.2.1 Wind energy

The developments in renewable electricity production in Germany have been very dynamic in wind energy. In absolute figures wind energy showed the strongest growth, reaching a combined generation potential of large and small hydropower installed capacity plants at the end of 2003 of 20.350 TWh. The actual generation of wind energy in 2003 was about 18,9 TWh due to a wind year that was 16% below average and to the fact that most wind turbines were installed at the end of that year. About 50% of the European wind energy capacity is installed in Germany.

2.2.2 Hydropower

Hydropower has the second-largest RES-E share, but it has not shown any significant development over the past five years (3,5% of total production).

2.2.3 Bioenergy

Biomass electricity, including the biodegradable fraction of municipal waste, is the third most important RES-E source, with about 6.970 TWh of electricity production in 2002.

2.2.4 Photovoltaics

Strong growth rates have also been achieved in the area of photovoltaics, reaching an installed capacity of 258 MW and a generation potential of about 190 GWh in 2002 and about 313 GWh in 2003. Table 3 shows electricity generation in Germany from RES for the years 1997 and 2002, as well as mean annual growth during this period.

Table 3: RES-electricity production (GWh) in Germany in 1997 and 2002

RES-E Technology	1997 [GWh]	2002 [GWh]	Mean annual growth [%]
Biogas	746	2.913	31%
Solid biomass	505	700	7%
Biowaste	1.168	2.035	12%
Geothermal electricity	0	0	
Hydro large-scale	11.696	16.340	7%

RES-E Technology	1997 [GWh]	2002 [GWh]	Mean annual growth [%]
Hydro small-scale	6.772	7.660	2%
Photovoltaics	27	176	45%
Wind onshore	3.034	17.200	41%
Total	**23.948**	**47.024**	**14%**
Share of total consumption [%]	4.50%	8.1%	

Source: *EC-ASEAN Energy Facility. European Experiences in Financing and Development of Renewable Energy Projects, Activity 8 Report.Denmark, May, 2006, p.30*

2.2.5 Heat sector

In the German heat sector, growth has been less rapid than in the electricity sector, although solar thermal collectors and heat pumps have attracted sizeable investment, especially from private households. A total collector area of about 5 million m^2 was installed by the end of 2002. Biomass heating is largely dominated by wood and wood-waste applications in households and a growing share of biogas, accounting for about 13% of biomass heat consumption by the end of 2001. The production of heat from wood in households has remained quite constant over recent years.

The biofuel sector has been growing very rapidly over the past 10 years, showing a doubling of production every two years. The existing biofuel mix is based almost entirely on biodiesel produced from rapeseed.[5]

2.3 Sweden

2.3.1 Bioenergy

The proportion of bioenergy used in the Swedish energy system has steadily increased from a little over 10% of the total energy supply in the 1980s to about 16% or 146-169 TWh in 2004. Most of the increase has been attributable to industry and district heating plants. The biofuels used in the Swedish energy system mainly consist of wood fuels, black liquors and tall oil pitches, and ethanol.

To a large extent, the expansion in biofuels has come about through an ambitious policy on renewable energy, and the Swedish Government is determined to continue pursuing this policy. Investment in bioenergy will contribute to a secure and sustainable energy supply, as well as growth and job creation.[6]

The leading domestic renewable source of energy is biofuels, i.e. fuels from the plant kingdom, which are mainly obtained from forest or farmland, although or-

---

[5] http://ec.europa.eu/energy/res/legislation/share_res_eu_en.htm.
[6] http://www.sweden.gov.se/sb/d/5745/a/19594.

ganic wastes from households and industry also make up an important share. In their various forms, biofuels can be used to produce electricity and heat, or as vehicle fuels.

The most important growth has occurred in the application of bioenergy, which grew by a factor of 2,5 in volume from 1990. Biomass produced 146-169 TWh in the year 2000, roughly two-thirds from wood and derived products, with the remainder from energy crops (agricultural fuels). However, this potential far exceeds the expected demand from the market. The majority of biomass, 5,1 Mtoe, is used directly, with the remainder being used to generate electricity and heat. Sweden and Finland rank in the top four for the use of wood energy. Their common point is the presence of vast expanses of wooded land on their territories (24 and 20 million hectares respectively) and a real tradition in exploiting wood for energy. A technique for producing electricity from wood is gasification, which is still in the development stage (EurObserv'ER, Wood Energy Barometer 2003).

Urban sewage plants are the main source of biogas, which is used as fuel or injected into the public natural gas network (particularly in Sweden). Sweden is the most advanced country in Europe when it comes to using biogas as fuel and has approximately 1.500 biogas-powered vehicles, mainly from professional fleets (EurObserv'ER, Biogas Barometer 2003).

2.3.2 Hydropower

Hydropower still remains the largest source of renewable energy in Sweden. In 2002 hydro generated 66 TWh, but 2003 was a very bad hydraulic year, with a total production of 53 TWh.[7]

Sweden's hydroelectric resources are mainly considered to be fully developed. However, questions of dam safety, improvements in efficiency, and the development of 'green', small-scale hydro are continuing to arise.[8]

Sweden is the only EU Member State where small hydropower (SHP) installed capacity has decreased, from 964 MW in 1990 to 935 MW in 2001 (-3%). About 3.807 GWh of electricity were generated by SHP in 2001, which represented about 5% of the electricity generated by hydropower and about 2% of the total electricity generated in Sweden. Hydropower contributed 50% of the total electricity generation of the country in 2001. In recent years, SHP is following a decreasing trend. Furthermore, public acceptance for building new hydropower in Sweden is extremely low, and considering this, the Swedish National Energy Administration estimates that potential will increase only 1 TWh from its 1996 value, when it reached its peak of 973 MW in 2005. Finally, the estimated installed capacity of small-scale hydropower plants in Sweden in 2000 was 1.062

---

[7] http://ec.europa.eu/energy/res/legislation/share_res_eu_en.htm.
[8] http://www.sgu.se/sgu/eng/samhalle/energi-klimat/fornybar-energi_info_e.html.

MW and the installed capacity of low temperature geothermal heat (excluding heat pumps) was 47,0 MWth with total energy production of 507 PJ (141 GWh). This is a minor increase from 1999, where the installed capacity was 1.050 MW.

2.3.3 Wind energy

Wind power started recently in Sweden (both onshore and offshore) and reached a level of around 0,6 TWh in 2002. By the end of 2003, the installed wind power capacity was 399 MW.[9]

Wind energy currently accounts for only a small share (less than 1 TWh) of the electricity produced in Sweden, but the sector is growing rapidly. The potential for wind energy is substantially larger. The expansion rate for wind energy has increased rapidly during the past few years. A national target has been set for creating the conditions for annual wind power production of 10 TWh by 2015.[10]

For environmental reasons, development of this energy source is increasingly shifting towards large offshore wind farms. A number of shallow waters and offshore banks within Sweden's territorial sea and exclusive economic zone are of interest for the establishment of such installations.

Table 4: RES electricity production (GWh) in Sweden in 1997 and 2002

RES-E Technology	1997 [GWh]	2002 [GWh][11]	Mean annual growth [%]
Biogas	46	17	-18
Solid Biomass	2.685	3.775	7
Biowaste	105	208	15
Geothermal electricity	0	0	
Hydro large-scale	64.560	62.370	-1
Hydro small-scale	2.582	3.630	7
Photovoltaics	0	0	0
Wind onshore	205	600	24
Total	**70.183**	**71.804**	**0,1**
Share of total consumption [%]	49,10%	46%	
Non-large hydro RES-E	5.623	8.230	

Source: Commission of the European Communities, Brussels, 26.5.2004, SEC (2004) 547, Commission Staff working document, The share of renewable energy in the EU,Country Profiles,Overview of Renewable Energy Sources in the Enlarged European Union, COM (2004) 366 final, p. 103

---

[9] http://ec.europa.eu/energy/res/legislation/share_res_eu_en.htm.
[10] http://www.sweden.gov.se/sb/d/5745/a/19594.
[11] Based on data by Swedish Energy Agency.

2.3.4 Heat pumps

In the heat sector the use of biomass, particularly district heating installations, has grown substantially over the past decade (by nearly 40% compared with 1990). The current use has reached a level of about 5 Mtoe.[12]

There are now over 300.000 ground source heat pumps in Sweden, mainly serving individual houses. The rock or soil normally delivers twice as much energy as needs to be supplied in the form of electricity. Gradually, the coefficient of performance (heat output per unit electricity) tends to decline, partly because insufficient energy is being transferred through the rock to the borehole, resulting in a fall in temperature in the hole and its immediate vicinity. In Sweden, a total of around 5 TWh of energy is extracted from rock and soil using this technology every year.[13]

In Sweden, the best potential for geothermal energy is considered to exist in areas where there are large bodies of groundwater at considerable depths (2-3 km), i.e. areas with thick layers of sedimentary bedrock or fault zones such as the Lake Vättern graben. Areas where meteorites have given rise to fractured bedrock at great depths are also judged to be of interest, including the Siljan Ring, the Dellen lakes and Björkö on Lake Mälaren. Sweden's largest commercial geothermal plant at present is in Lund. Drawing water at 20°C from sedimentary strata at a depth of some 700 m, the plant meets 30% of the city's district heating needs (i.e. 250 GWh). The heat contained in the water is heat-exchanged to achieve the required temperature.

2.3.5 Fuel peat

Sweden has one of the highest proportions of peatlands in the world, with peat covering some 15% of its surface. Peat for use as a fuel is currently extracted from around 12.000 ha of the country's roughly 6,4 million ha of peatlands. The energy content of the country's peat resources (on a dry matter basis) is estimated at around 57.000 TWh.

2.3.6 Solar energy

Solar thermal collectors have been introduced in Sweden, but their contribution still remains small. The market for solar thermal applications grew by 7% in 2002 to nearly 0,2 million m² installed capacity.[14] The demonstration plant in Kungälv is currently the largest solar heating plant in Europe (March 2001).

---

[12] http://ec.europa.eu/energy/res/legislation/share_res_eu_en.htm.
[13] http://ec.europa.eu/energy/res/legislation/share_res_eu_en.htm.
[14] http://ec.europa.eu/energy/res/legislation/share_res_eu_en.htm.

The photovoltaic (PV) power installed during 2002 in Sweden amounted to 265 kW, which is a slight increase compared with the past few years. This means, however, that the trend for the cumulative installed power is still best described as a linear increase. The exponential increase that can be observed when analysing the installed power in all the IEA countries is not present in Sweden. This is probably due to the lack of government subsidies and long-term PV goals.

3. Renewable Energy Policy Outlook

The EU Energy Acquis consists of rules and policies, notably regarding competition and state aids (including in the coal sector), the internal energy market (opening up of the electricity and gas markets, promotion of renewable energy sources), energy efficiency, nuclear energy and nuclear safety and radiation protection.[15]

EU energy policy objectives include:

- Improvement of competitiveness;
- security of energy supplies;
- protection of the environment.

In March 2007, European leaders signed a binding EU-wide target to source 20% of their energy needs from renewables such as biomass, hydro, wind and solar power by 2020. On 23 January 2008, the Commission put forward differentiated targets for each EU Member State, based on the per capita GDP of each country.

March 2007: EU summit endorses Commission roadmap with:

- A binding target to have 20% of the EU's overall energy consumption coming from renewables by 2020;
- as part of the overall target, a binding minimum target for each Member State to achieve at least 10% of their transport fuel consumption from biofuels.

However, the binding character of this target is "subject to production being sustainable" and to "second-generation biofuels becoming commercially available".

January 2008: EU Commission presents a proposal for a directive to reach the targets set in March. Proposal forwarded to the EU Council and Parliament for approval. Agreement on 9 December 2008. According to this new legislation, each EU country will be required to significantly increase the contribution of renewable energies to its energy mix, leading to an overall EU share of 20% by

---

[15] http://www.europarl.europa.eu/meetdocs/2004_2009/documents/fd/d-tr20060425_06/d-tr 20060425_06en.pdf.

2020. A 10% share of biofuels in transport by 2020, part of the overall 20% renewables target, was agreed previously under the condition that indirect land-use considerations and other sustainability criteria be taken into account

1st half 2009: Target date for the adoption of the legislation.

31 March 2010: Deadline for EU states to present National Action Plans (NAPs) on renewables.

- Renewable energies such as wind power, solar energy, hydropower and biomass can play a major role in tackling the twin challenge of energy security and global warming because they are not depletable and produce less greenhouse gas emissions than fossil fuels;
- since the energy crises of the 1970s, a number of industrial nations have launched programmes to develop renewable energy solutions, but the return of low oil prices has prevented renewable energies from picking up on a large commercial scale;
- renewable energies today cover 13,1% of global primary energy supply and 17,9% of global electricity production (IEA 2007). The IEA's World Energy Outlook 2006 foresees in its Alternative Policy Scenario that the share of renewables in global energy consumption by 2030 will remain largely unchanged at 14%. Renewables in electricity generation are expected to grow to around 25%, according to the IEA.[16]

3.1 Sweden

Sweden's National Programme for Energy Efficiency and Energy-smart Construction (Govt Bill 2005/06:145) proposes an Energy Declaration of Buildings Act to harmonise domestic legislation with the EPBD. Under Swedish legislation, buildings will be subject to inspections, and certain information about a building's energy use and indoor environment will be certified in an energy declaration when buildings are constructed, sold or rented out. The building owner will be able to reduce the costs of energy use through the measures proposed in the energy declaration.

Sweden's energy policy, as decided by the Swedish Parliament in 1997, provides secure short-term supplies of electricity or other energy on competitive terms. The country's energy policy is intended to create conditions for efficient use and cost-efficient supply of energy, with minimum adverse effects on health, the environment and climate, while at the same time assisting the move towards an ecologically sustainable society.

---

[16] http://www.euractiv.com/en/energy/eu-renewable-energy-policy/article-117536.

The Government programme running since July 1997 supports renewable energy investments in order to encourage increased production of renewable electricity, particularly from biomass and wind. The Parliamentary decision on energy policy in June 1997 [Toward a Sustainable Energy Supply] included a strategy for reducing the energy sector's impact on climate. The strategy is based on the view that successful international cooperation requires an equitable distribution of commitments and mitigation costs, and that national circumstances should be taken into account when determining environmental commitments.

Grants available since July 1997 are:

- 25% for investments in CHP (Combined Heat and Power) plants based on biomass (up to 3.000 SEK/kWe), with a 5-year budget of MSEK 450;
- 15% for wind turbines over 200 kW, with a 5-year budget of MSEK 300;
- 5% for environmentally friendly, small-scale (<1.5 MW) hydro plants, with a 5-year budget of MSEK 150.

This compares to similar grants for CHP and 35% grants for wind turbines >60 kW under the previous investment support programme, initiated in July 1991. In addition to the 1997 investment support programme, the government set up a 5-year technology procurement programme for renewable electricity production from January 1998. Total funds for the procurement programme are MSEK 100.

A seven-year programme aimed at devising an ecologically sustainable energy system was initiated in January 1998.

Total programme funds of SEK 5,28 billion (on average in 1997, 1 US $ = 7.635 SEK) were made available over the seven-year period, including SEK 2,73 billion on energy research in Sweden. This reverses the previous downward trend in Government R&D expenditure. An additional SEK 1,61 billion is dedicated to the support of commercial electricity production from renewables.

A new public authority, the Swedish National Energy Administration (STEM), was set up on 1 January 1998, with the responsibility for implementing most of the energy policy programmes and coordinating the work of restructuring the energy system. In addition, STEM is also responsible for monitoring developments in the energy and environmental fields and for providing information on the current energy situation. This covers aspects such as changes in the structure and pattern of energy supply and use, energy prices and energy taxes and the effects of the energy system on the environment.

The objective of the Swedish Government's energy policy is to secure a reliable supply of electricity and other forms of energy at internationally competitive prices, both for the short and the long term. Sweden has decided that an energy policy should create conditions for efficient and sustainable energy use, as well as a cost-effective Swedish energy supply with minimum negative impact on health, the environment and the climate. It should also facilitate the transition to an ecologically sustainable society. To achieve this, global cooperation is required.[17]

The vision in Swedish energy policy is that the country will obtain all its energy from renewable energy sources in the long term. The most important policy instrument in promoting renewable electricity production is the electricity certificate system that was introduced in 2003.[18]

The latest policy paper was adopted by the Supreme Planning Council on 17 March 2004. The main objectives are to complete privatisation of state-owned companies, liberalisation of markets with all of its elements, diversification of resources (introduction of renewables and nuclear) and to decrease dependence on imported fuels.

The Renewable Electricity Law was adopted in 2005, as the transposition of Directive 2001/77/EC within the scope adoption of EU Acquis. The Law, which enables the Government to purchase the electricity from renewable energy sources to a maximum of 20%, was fully operational as of 2007.

The Draft Law on Energy Efficiency as well as the Draft Law on Geothermal Resources are currently being discussed in the Swedish Parliament. Biodiesel and bioethanol have been developed under the Petrol Markets and Tobacco Markets Law, respectively.

The Ministry of Energy and Natural Resources General Directorate of Energy Affairs is the key government institution in planning.[19]

3.1.1 Main supporting policies

Swedish RES-E policy is composed of the following mechanisms:

Tradable Green Certificates. These were introduced in 2003. The Renewable Energy with Green Certificates Bill that came into force on 1 January 2007 shifts the quota obligation from electricity users to electricity suppliers, and incorporates a new target of 17 TWh by 2016. This system is a market-based support system to assist the expansion of electricity production in Sweden from renewable energy sources and peat. One electricity certificate unit is issued to each approved producer for each produced and metered megawatt hour of electricity from renewable energy sources, or from peat. Plants commissioned after the start of the electricity certificate system are entitled to receive electricity certificates for 15 years, or until the end of 2030, whichever is the earlier. Plants that were started up before the certificate system was introduced are entitled to certificates until the end of 2012.

---

[17] http://www.regeringen.se/sb/d/5745/a/19594.
[18] http://www.regeringen.se/content/1/c6/06/47/22/2c000830.pdf.
[19] http://www.rec.org/REEEP/energy_country_profiles/turkey.pdf.

Plants that at the time of their construction or conversion received a public investment grant after 15 February 1998 are entitled to certificates until the end of 2014.

Figure 3: Number of electricity certificates issued in Sweden, by type of energy source, 2003-2006

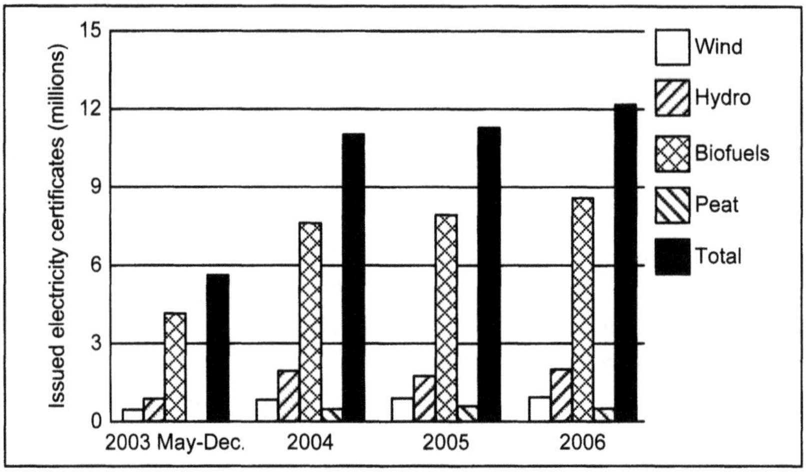

Source: Svenska Kraftnäts accounting system, Cesar

The demand for electricity certificates is created by the obligation that electricity suppliers and certain electricity users are required to purchase certificates corresponding to a particular proportion of their electricity sales and/or electricity use. This proportion, or quota, has been decided for each calendar year, and set at such a value that the system will play its part in achieving the objective of increasing the production of electricity from renewable energy sources by 17 TWh by 2016, relative to production in 2002. Grants are available for local authority land use planning for wind power. The electricity certificate system supports renewable electricity. As the country is experiencing some difficulties in seeking to achieve its planned objective for wind power (namely 10 TWh of wind power production by 2015), this grant seeks to assist achievement of the planning objective by supporting local authority land use planning for 2007 and 2008.

Premium Tariff. The environmental premium tariff for wind power is a transitory measure and will be progressively phased out by 2009 for onshore wind.[20]

Tax exemption. Sweden has also introduced energy taxes, which were originally intended to stimulate energy saving and renewable energy, and are now also intended to reduce the environmental impact of energy. These energy taxes consist of a general energy tax, a CO_2 tax and a sulphur tax. In 1991, this general energy

---

[20] http://www.erec.org/fileadmin/erec_docs/Projcet_Documents/RES2020/SWEDEN_RES_Policy_Review_April_2008.pdf.

tax on fuels and electricity was halved, to make way for a new CO_2 tax. In 1992, the Government also introduced a NOX tax. Sweden is implementing a 'green tax exchange' whereby taxes on environmentally harmful activities are raised, while taxes on labour are reduced by a roughly equal amount. A step in the green tax exchange was performed in 2001, when taxation on energy products was increased by about SEK 3 billion. The CO_2 tax rate was raised from SEK 370 per tonne to SEK 530 per tonne. The energy tax on diesel went up by SEK 0.1 per litre and the energy tax on electricity was raised by SEK 0,018 per kWh. In accordance with the principles of the green tax exchange, the bulk of this increase was offset by a higher tax-free allowance and a reduction in employer's levies. A further green tax exchange was carried out in 2002, when taxation on energy products was raised by about SEK 1,7 billion. The tax rate on CO_2 was raised from SEK 530 per tonne to SEK 630 per tonne, and the energy tax on electricity went up by SEK 0,012 per kWh; taxes on labour were reduced by a compensatory amount. The rises in the tax on CO_2 and electricity affect only consumers. Taxes on the transport sector have been left largely unchanged. The reductions in CO_2 tax that apply to the industries with exemptions (i.e. manufacturing, agriculture, forestry and aquaculture) have been adjusted from 50% to 70%. This adjustment largely offsets the higher CO_2 tax and keeps the overall tax position of these sectors unchanged (IEA Standard Review Sweden 2002).

Today only 8% of Swedish houses are heated by oil, and those households get tax rebates if they switch to renewable sources. Renewable heat has been supported in an indirect way by raising taxes on fossil fuels.

Green taxes such as the CO_2 tax promote biofuels in an indirect way. In addition, the Swedish Government is currently increasing the number of alternative fuel pumps and it ensured that 36% of the vehicles it used in 2006 were fuelled, either wholly or in part, by biogas, ethanol or electricity. Finally, a subsidy has been granted for investment in filling stations for biogas and other renewable fuels: MSEK 150 was set aside in 2006 and 2007. Bioethanol and biogas are exempted from the tax on petroleum products and from the tax on CO_2. There are also tax exemptions for large service stations that supply more than 3.000 m³ of petrol or diesel a year if they convert to either ethanol or biogas. Since an ethanol filling station is much cheaper than a biogas filling station, ethanol filling stations are normally the preferred choice by the station owners.[21]

Energy efficiency promotion. This Act aims to promote energy efficiency and a good indoor environment in buildings. Under proposed transitional provisions, energy declarations for premises used for public activities (known as special buildings) and multi-dwelling buildings (apartment blocks) must be carried out by

---

[21] http://www.energy.eu/renewables/factsheets/2008_res_sheet_sweden_en.pdf.

the end of 2008. Energy declaration of other buildings will begin on 1 January 2009.[22]

Support for purchase of energy-efficient windows and biomass boilers is set at 30% of the cost exceeding SEK 10.000; support capped at SEK 15.000. One-family households are eligible for these tax credits.[23]

National RES targets. Sweden has set a target of a 10 TWh increase for RES electricity between 2002 and 2010. According to the analysis, the increase will be 10,9 TWh. Including existing RES electricity production, Sweden's total production of RES electricity will be 80.4 TWh in 2010. This corresponds to 50,5% of gross national electricity consumption.[24]

During the 1990s, the Swedish electricity market was reformed in several steps. Since 1 January 1996, Sweden has a liberalised electricity market. All consumers are free to choose their electricity supplier. The objectives of the reform have been to increase the freedom of choice for electricity consumers and to create conditions for greater pressure on prices and costs in the electricity supply.

The electricity networks throughout the country must be open to all players on the market who have paid a connection charge somewhere in the country. The Swedish National Energy Administration is responsible for monitoring the network operations on the reformed electricity market. The Swedish electricity system is connected with the electricity systems of other Nordic countries (excl. Iceland). As a result of the reformation process in the Nordic countries, there is nowadays free access to the interconnections between Sweden, Finland and Norway. Thus, it is possible for Swedish, Norwegian and Finnish power generators and consumers to buy and sell electricity in the three countries. Major Danish companies are also active on the Nordic electricity market, as well as companies from other neighbouring countries.

Buy-back rates. In Sweden, electricity produced from renewables is supported by buy-back rates. The buy-back rates for 2002 for electricity produced in small hydropower plants and sold to the utilities or traders consisted of the following:

- Price for electricity sold to an utility or a trader working on the deregulated market;

---

[22] http://www.erec.org/fileadmin/erec_docs/Projcet_Documents/RES2020/SWEDEN_RES_Policy_Review_April_2008.pdf.

[23] http://www.iea.org/textbase/pm/?mode=re&id=2547&action=detail.

[24] Analysis of Sweden's success in achieving its national indicative targets for RES electricity. Drawn up pursuant to Article 3(3) of Directive 2001/77/EC of the European Parliament and of the Council on the promotion of electricity produced from renewable energy sources in the internal electricity market, 2003.

- price given by the local grid owner that is equivalent to the cost reduction in the net due to the locally produced power;
- state support at a fixed price per kWh.

The Feed-in system (with support prices in the range 0,97 to 1,95 € cents/kWh) was the main instrument used to promote RES-E until 2003. As from 1 May 2003, a TGC scheme is being implemented in order to reach the ambitious targets set by the Government (to increase RES-E consumption by 10 TWh from 2002 to 2010). TGCs apply to wind, solar, geothermal, hydro-electric, wave and biofuel power. The framework is a quota-based system, meaning that an obligation is placed on all electricity consumers to purchase an increasing proportion of their electricity consumption from renewable sources, starting from 7,4% in 2003 and reaching 16,9% in 2010. Consumers fulfil this obligation by buying enough certificates from green generators to cover such a quota and surrendering them to the Swedish Energy Agency. In practice, electricity-distributing companies will take care of the obligation on behalf of any clients who do not give notice that they wish to manage their quota obligation themselves. Energy-intensive industries are exempted from the obligation in the initial phases of the scheme. Whether or not they will have an obligation in the future is under consideration. Any user/supplier who has surplus electricity certificates may sell them or save them for the needs of future years, since green certificates have an unlimited life. The certificate price will be set on the market.

However, there is a minimum price and a theoretical maximum price indicated by the penalty charge. The minimum price is the buy-out price at which the Swedish Energy Agency has to buy the certificates from the producers if they find no buyers for their certificates. This minimum price starts at 6,5 €/MWh in 2003 and will be thereafter gradually reduced and entirely phased out in 2008. There is also a penalty charge for those electricity consumers who do not fulfil their quota obligation by showing enough certificates. The penalty charge is 150% of the average certificate price during the year, but with a maximum of 19,1 €/MWh for certificates to be surrendered during 2004 and 26,2 €/MWh for certificates for 2005 (plus taxes). The green certificates market is still extremely thin since most deals are closed bilaterally between the green generators and the buyers. There are two types of contracts in the market: (a) the *fixed volume contract* in which the buyer is guaranteed a certain amount of electricity and (b) the *falling volume contract* based upon an estimation of how much the green generator expects to produce. The former is usually slightly more expensive.

Since it is doubtful whether wind power is going be deployed on a large scale under the relatively low level of TGC prices in Sweden, the Swedish government has proposed transitional subsidies for wind power production. This bonus will be given until a windmill has run for 25.000 equivalent full load hours from it started to produce power. This subsidy will only be available for a five-year transitional period 2003-2007, in which the bonus will gradually be phased out.

3.1.2 Transport

Biogas as a motor fuel is mainly exempted from tax. The Swedish Government has the right to apply exemptions or reductions in the rates of duty to fuels used in the field of pilot projects for the technological development of more environmentally friendly products and in particular in relation to fuels from renewable resources. With reference to this regulation, the Government has given relief from excise duties to pure ethyl alcohol (ethanol) used as motor fuels in pilot projects and has set the energy tax rate for 1998 at 0,9 SEK (0,1 ECU) per litre Furthermore, the Government has granted tax relief for rapeseed methyl ester (RME). Electrically powered rail transport is energy-efficient and has a very small impact on the environment. This was one of the reasons why the Government adopted a ten-year investment plan for railway infrastructure amounting to 36 billion SEK in 1998. A new authority, *Rikstrafiken,* has been created to promote the long-distance public transport system. It will encourage the use of public transport and support unprofitable public transport considered socially desirable.

Since 1991, the tax on petrol includes a carbon dioxide tax estimated to have generated about 11% of the state's revenues from road traffic-related taxes in 1996. VAT of 23,46% was imposed on petrol in 1990, and has since been raised to 25%. Up until October 1993, a tax was paid on diesel-powered trucks, cars and buses, based on distance driven (kilometre tax). This tax required border controls and was replaced by a diesel oil tax after Sweden joined the European Union. Sweden also levies sales taxes, annual vehicle taxes and user charges. Sales taxes are levied on light vehicles only. The annual vehicle tax is differentiated according to vehicle weight and fuel. Only heavy goods vehicles are charged so-called user charges (Eurovignette).

Rail transport is excluded from energy taxes, but there is a user fee system. Parliament reduced this fee in 1998 in order to increase the relative competitiveness of rail transport. Energy taxes do not apply to maritime and air transport. Shipping pays environmentally differentiated seaway charges, and aviation pays route charges according to the Eurocontrol procedure (IEA Energy Efficiency Update 2003 Sweden).

3.1.3 Research and development

In 2000, the Swedish Government devoted about SEK 646 million to energy-related R&D, making its energy R&D budget one of the largest in Europe. Of the total budget, 36% was spent on conservation, 34% on renewable energy technologies, and 10% on power and storage systems (IEA Standard Review Sweden 2002).

Swedish research and development actively supports technological developments in renewable energy. Biomass research, development and demonstration receive total funding of about SEK 400 million (EUR 35 million) per year from the Government. Electricity companies and other industries also provide funds. The main areas of support are combustion and conversion technologies, demonstration of pre-competitive technologies, fuel production, harvesting supply programmes and ash recycling.

3.2 Turkey

The Turkish Government has developed an energy policy aimed at diversifying energy sources and suppliers and attracting private capital. Special attention in the Government's energy policy is paid to the development of international cooperation. Turkey has developed and implemented several energy efficiency projects, aiming to increase energy efficiency in the industry, transport and residential sectors. Analyses of the relationships among the primary energy consumption, Gross Domestic Product (GDP), population, electricity production and consumption in Turkey show that during the past 10 years, while population increased by 1,92%, GDP and Total Primary Energy Consumption increased by 3,03 and 3,38, respectively. Net Electricity Consumption, Net Electricity Production and Gross Electricity Production increased by about 10% during the same period.[25]

3.2.1 Energy policy in Turkey

The real beginning for renewable energy policy was the definition of renewable energy sources in the decree on Modification of the Licence Regulation in the Electricity Market in 2003. Before then, there was no national renewable energy policy and few government incentives existed to promote market deployment of renewable energy. However, the Electricity Market Licensing Regulation, in itself, is not expected to be sufficient to overcome the high investment cost, risk and lack of security associated with the entrance of renewable power plants into the electricity market.[26]

The industrial sector accounted for 40% of total final energy consumption and for 54% of electricity consumption in 2000, while the agriculture, household and services sectors together accounted for 40% of final energy consumption and 46% of electricity consumption. Although all four sectors have important potential for energy conservation, industry has been targeted as a priority area for energy conservation programmes owing to the projected rapid expansion of industrial energy de-

---

[25] http://www.planbleu.org/publications/atelier_energie/TR_National_Study_Final.pdf.
[26] http://www.planbleu.org/publications/atelier_energie/TR_Summary.pdf.

mand. On the other hand, the structure of the Turkey's industry is energy-intensive.

Turkey is to be the recipient of a US$ 202 million renewable energy loan provided by the World Bank to be disbursed as loans via financial intermediaries to interested investors in building renewable energy-sourced electricity generation. These loans are expected to finance 30-40% of associated capital costs. The aim of the Renewable Energy Programme is to increase privately owned and operated power generation from renewables sources within a market-based framework, which is being implemented according to the Electricity Market Law and the Electricity Sector Reform Strategy. This programme will assist the Directorate of the Ministry of Energy and Natural Resources (MENR) in the preparation of a renewable energy law, and in defining the required changes and modifications related to legislation such as the Electricity Market Law to better accommodate greater private sector involvement.

3.2.2 Energy efficiency

In Turkey, although energy consumption per capita is one-quarter of the average for the OECD countries, energy intensity value is twice as much as that of these countries. According to the International Energy Agency (IEA) database, the OECD average of energy intensity is 0,19, but is 0,09 for Japan and 0,38 for Turkey. These data indicate that Turkey uses energy less than others and in an inefficient way. With the economical development, Turkey should not only increase its energy usage per capita but also decrease its energy intensity. This can be only realised by means of energy efficiency studies. Considering current projections for 2020, it can be said that the primary energy consumption will be about 222 Mtoe and there will be a 15% energy saving potential.[27]

With regard to the 'Energy Efficiency Strategy' adopted in 2004 by the Ministry of Energy and Natural Resources (MENR) with the aim of improving end-use energy efficiency in industry, building and transportation sectors, Turkey will have an opportunity to get the stated energy saving potentials for the related sectors by regulatory arrangements on the basis of European Union (EU) Acquis. To this end, the Law on Energy Efficiency was enacted and published in May 2007. This Law, comprising required legal frameworks of improving efficiency in production, transmission, distribution and consumption steps of the energy cycle, can be evaluated as very comprehensive within the current frame. In the case of e.g. lighting, the proposed implementations relate to the end-use sectors, namely industry and building and transportation.

Since enactment of the Law in May 2007, there have been very intensive studies on energy efficiency in Turkey, especially on the topic of public awareness.

---

[27] http://www.balkanlight.eu/abstracts_pdf/b13.pdf.

Studies, various campaigns, conferences and seminars are being organised with the help of non-governmental organisations.

3.2.3 Market deployment

Market deployment policies for renewables started in 1984 with third-party financing and excise and sales tax exemptions. Capital grants were offered in 2001. The Turkish Government's approach to the deployment of renewables reflects its priorities to develop indigenous and renewable resources in conjunction with the expansion of privately owned and operated power generation from renewable sources.[28]

The build-own-transfer (BOT) and the build-own-operate (BOO) schemes were put in place in 1984 and financed major power projects (not limited to renewables) with the main objective of attracting private investors. BOT projects were granted a treasury guarantee. Although these BOT and BOO approaches attracted significant investment, they also created large contingent public obligations, with the Government covering the market risk through take-or-pay contracts. The economic crisis of 2000 and pressure from the International Monetary Fund, however, brought an end to the treasury guarantees, except for the 29 BOT projects with contracts already in place.

In October 2005, the EU opened accession negotiations with Turkey. The first phase of the accession process, the analytical examination of the Acquis (screening), was completed in October 2006.[29]

In the 2006 Report of the Commission, progress on the adoption of the Community Acquis was judged to have been reached in following areas:

Security of supply: Turkey has already introduced major measures and its oil reserves are more or less at the 90 day level required by the Acquis. Turkey also has an important role to play in the EU's security of supply, since it is a transit country for oil and gas from the Caspian Sea, the Black Sea and Central Asia. The construction of a Turkey-Greece gas interconnector started in July 2005 and aimed to be operational in spring 2007. Turkey also supports the *Nabucco* gas pipeline project. Construction of this 3.300-kilometre pipeline began in 2008 and is planned to be finished in 2011. A new South Caucasus gas pipeline (Baku-Tibilis-Erzurum) became operational in 2006.

Internal energy market: Some progress can be reported. The privatisation process of distribution assets has started for three regions. Implementing regulations have been enacted on electricity demand forecasting and cross-border electricity

---

[28] Global warming and renewable energy sources for sustainable development: A case study in Turkey: Selçuk Bilgen, Sedat Keleş, Abdullah Kaygusuz, Ahmet Sarı and Kamil Kaygusuz 2006.

[29] http://www.eva.ac.at/enercee/enlargement.htm.

trade. The threshold for eligible consumers has been reduced to 6 GWh. A new amendment, however, allows cross-subsidies and vertical integration. High electricity losses, including theft, have persisted. Unchanged electricity tariffs in the context of rising gas import prices may in the short term result in real capacity reductions. Turkey is not yet a member of the Union for the Coordination of Transmission of Energy and has not signed the Energy Community Treaty establishing a regional energy market in southeast Europe.

Energy efficiency: In the progress report of 2006, the EU Commission notes that in this section no progress can be reported, since Turkey still does not have a framework law for its promotion.

Renewable energies: Some progress has been made on renewable energy sources. However, Turkey has not set itself an ambitious target yet for their increase. An implementing regulation on the guarantee of origin has been issued. Turkey is partially aligned in this area.

As regards renewable energy sources, some progress was noted by the European Commission in the 2005 progress report. A Law on the Use of Renewable Energy Sources in Electricity Generation was adopted in May 2005, establishing the necessary legal framework for the promotion of renewable energy. The law provides transitional arrangements (until 2011) for more competitive prices for electricity generated from plants that have a renewable energy resource certificate, and other incentives for investments in renewables. However, the Law does not set a target for electricity generated from renewable sources by 2010, as foreseen by the relevant EU directive. Given Turkey's significant untapped potential for renewable energy sources, it should set itself an ambitious target for their further development, including geothermal energy. Turkey would be recommended to develop an overall strategy for renewable energy sources.

Nuclear energy: Turkey currently does not operate any commercial nuclear power plants, but has announced plans to promote the construction of nuclear power capacity of 5.000 MW by 2020. The independence of the Turkish Atomic Energy Authority (TAEK) needs attention. Supervisory responsibilities are not separated from research and the promotion of nuclear energy.

Nuclear safety and radiation protection: No new implementing regulation has been enacted. Substantial upgrading of existing facilities will be needed, including radioactive waste management and storage facilities. Turkey has not acceded to the Joint Convention on the Safety of Spent Fuel Management and on the Safety of Radioactive Waste Management, to which Euratom became a contracting party in January 2006.

Energy Efficiency Law: This Law was adopted on 18 April 2007 with the aim of increasing efficiency in using energy sources and energy in order to use energy effectively, avoid waste, ease the burden of energy costs on the economy and protect the environment. This law covers principles and procedures applicable to increasing and promoting energy efficiency in energy generation, transmission, distribution and consumption phases at industrial establishments, buildings, power generation plants, transmission and distribution networks and transport, raising energy awareness in the general public, and utilising renewable energy sources. Implementations relating to supporting energy efficiency implementation projects, reducing energy intensity, and research and development projects must be carried out according to prescribed principles and procedures.

Law on Utilisation of Renewable Energy Resources for the Purpose of Generating Electrical Energy:[30] The purpose of this Law is to expand the utilisation of renewable energy resources for generating electrical energy, to benefit from these resources in secure, economic and qualified manner, to increase the diversification of energy resources, to reduce greenhouse gas emissions, to assess waste products, to protect the environment and to develop the related manufacturing sector for realising these objectives. This Law encompasses the procedures and principles for conservation of the renewable energy resource areas, certification of the energy generated from these resources and utilisation of these resources.

Renewable Energy Resource Certificate: Any legal entity holding a generation licence must be issued by the EMRA with a Renewable Energy Resource Certificate (RES Certificate) for the purpose of identification and monitoring of the resource type in purchasing and sale of the electrical energy generated from renewable energy resources in the domestic and international markets.

3.3 Germany

Renewable energy policy in Germany began in 1974, after the first oil crisis. For about a decade and a half, this policy consisted almost exclusively of the promotion of research (from training personnel to development of prototypes and laboratory production).

Since 1979, there were also initial efforts to stimulate demand for RES-E by the use of tariffs. At that time the Government relied on the national competition law to oblige electricity distributors to purchase electricity from renewable sources produced in their area of supply based on the principle of avoided costs.

The accident in Chernobyl in 1986 had a deep impact in Germany. Public opinion had been divided about evenly on the question of nuclear power between 1976 and 1985. This changed dramatically in 1986. Within two years, opposition to nuclear power increased to over 70%, while support barely exceeded 10% (Jahn 1992).

---

[30] Law on Utilization of Renewable Energy Resources for the Purpose of Generating Electrical Energy, Law No. 5346, Ratification Date: 10.05.2005, Enactment Date: 18.05.2005. SECTION ONE. Purpose, Scope, Definitions and Abbreviations.

In the late 1980s, several measures were adopted to create markets for RES-E technologies. These were in particular the 100/250 MW wind programme and the 1.000 solar roofs programme. When in 1988 two backbench conservative MPs in the Bundestag proposed a feed-in tariff to support wind energy, the Government, to buy off the dissenters, initiated two important market creation programmes for RES-E: a 100 MW wind programme and 1.000 roofs programme for photovoltaic (Kords 1993). From 1991 to 1995, under the 1000 roofs programme, applicants received 50% funding of investment costs from the Federal Government plus 20% from the Land Government (in the new *Länder* the percentage was 60 and 10% respectively). Eventually 2250 roofs were equipped with PV modules, leading to about five MW of installations (Ristau 1998; Staiss 2000: I-140). As to wind energy, a programme for subsidising 100 MW (subsequently 250 MW) of wind turbines (by a payment of € 0,04/kWh, later reduced to € 0,03) was introduced, motivated by the need to gain practical experience with different approaches under real-life conditions. In addition, the legal framework for electricity tariffs was modified in 1989 in such a way as to allow compensation of RES-E generators above the level of avoided costs, a provision that was to play a role in the 1990s for the development of photovoltaics.

The 1990 Feed-In Law. A federal Electricity Feed-In Law (StrEG) was adopted in 1991 and became the most important instrument for the promotion of renewable energy in Germany during the 1990s. It obligated public utilities to purchase renewably generated power from wind, solar, hydro, biomass and landfill gas sources, on a yearly fixed rate basis, based on the utilities' average revenue per kWh. Remuneration to wind producers was set at 90% of the average retail electricity rate, while for other renewable power providers, compensation was set at 65-80%, depending on plant size, with smaller plants receiving the higher subsidy level. The StrEG effectively subsidised the operation of commercial wind installations at 4,1 € cents/kWh, and jump-started the breakthrough of the wind power market in the 1990s, as illustrated in Table 2. In addition, investment in wind power installations was subsidised by a domestic, state-owned development bank, the Deutsche Ausgleichsbank, which offered low-interest, government-guaranteed loans for new wind power development.

3.3.1 Photovoltaics

While the Feed-In Law of 1990, combined with the 250 MW wind programme, led to the breakthrough for wind, photovoltaics did not benefit similarly. The 1.000 roofs programme of 1989 had been a success and led to installations of 5,3 MW by 1993, but this market volume did not justify the installation of new production facilities in the solar cell industry. The Feed-In Law provided little help since rates did not come near PV costs, and a new demonstration programme was not forthcoming.

The 1989 modification of the federal framework regulation on electricity tariffs mentioned above permitted utilities to conclude cost-covering contracts for electricity using renewable energy technologies, even if these 'full cost rates' exceeded the long-term avoided costs of the utilities concerned.

As the process first started in Aachen, this is known as the Aachen model (Solarförderverein 2002; Staiss & Räuber 2002).[31] Additional help came from several *Länder* market introduction programmes, most strongly in North Rhine-Westphalia. Some states acted through their utilities, subsidising solar installations for special purposes, e.g. schools (Bayernwerk in Bavaria, or BEWAG in Berlin). Some offered 'cost-orientated rates' somewhat below the level of full cost rates (thus HEW in Hamburg). Finally, Greenpeace gathered several thousand orders for solar cell rooftop 'Cyrus installations' (Ristau 1998). Due to these initiatives, the market did not collapse at the end of the 1000 roof programme but continued to grow, attracting new firms and demonstrating public support for PV.

At the same time, some large German PV firms moved their production to the United States; thus increasing pressure on public authorities to come up with new support measures. On the promise of such a programme, ASE (one of the large cell manufacturers) invested in a new plant in Germany, which started production in 1998 with a capacity of 20 MW. Similar promises induced Shell to enter the German solar cell industry with a 9,5 MW plant in Gelsenkirchen in the same year (Jacobsson & Lauber, in press). These activities, new organisations and investments increased the pressure for market creation by the Government.

3.3.2 Reforms

German electricity regulation traditionally relied on a mix of public and private law. Basic energy law was embodied in the Energy Supply Industry Act (*Energiewirtschaftsgesetz*) adopted in December 1935 and laying down the framework conditions for a cheap and secure electricity supply. It defined German state control of the sector for more than 60 years. The other important piece of legislation is the Monopolies Act, which generally exempted electricity supply. Contracts for concessions, territorial boundaries, supply to special customers, the technical conditions for feeding surplus electricity into the grid, reserve deliveries and other arrangements are all based on private law.

On 28 September 1999, the German Government, the SPD and Green Parliamentary Parties and leading Unionists signed a common statement confirming the basic principles of the energy law reforms, namely the end of demarcation

---

[31] In 1989 Bayernwerk introduced the first 'green pricing' scheme, which involved investment in a 50 kW$_p$ plant. Many such schemes followed, for instance by RWE in 1996. About 15.000 subscribers eventually paid an eco-tariff (twice the normal tariff) for electricity generated by solar cells, hydropower and wind (Jacobsson & Lauber, in press).

treaties, full opening of the network to all suppliers and free choice of supplier for all customer groups (ARE 2000[12]). Liberalisation made a little more headway in 2003 and 2004.

3.3.3 Climate change policy

Within the framework of the Kyoto Protocol and the European burden-sharing concept, Germany pledged to reduce GHG emissions by 21% from 1990 to 2008/12. In addition, in 1995 the Government had pledged a 25% reduction in CO_2 emissions by 2005. By 2000, a reduction of about 18 to 20%, corresponding to 180 to 200 million tons of CO_2, was already achieved, so that the shortfall amounted to 50 to 70 million tons of additional reduction. This was to be achieved by the Government's Climate Change Policy Action Programme of October 2000. Both RESA and the CHP Act are integral parts of this programme. These two areas of activity are expected to contribute reductions of 15 Mt CO_2 and 23 Mt CO_2 respectively, or about 50% of the target (Bundesregierung 2000: 9, 77, 80).

Eco-Tax Reform: The 1999 Ecological Tax Reform (ETR) initially increased the taxes on motor fuels, fuel oils, and natural gas, and also levied an electricity tax across all sectors. These taxes have increased in subsequent years. The ETR has helped to broaden the use of biofuels, which are exempt from taxation under German law, but has had a neutral effect on wind, solar, and other sources of renewable electricity, since all electricity providers are subject to the ETR power levy.

Renewable energy: The German Government formulated a target to increase the share of RES-E in the electricity supply to 12,5% in 2010 and 50% in 2050; in 2004 the goal of 20% by 2020 was added. The long-term target must be viewed as a programmatic goal, which in concert with energy efficiency programmes is ambitious but not unrealistic either technically or economically.

Several measures were taken in favour of renewable energy. They included a five-year market incentive programme for RES which provided about € 445 million from 1999 to 2002. A tax break on biofuels was applied in keeping with an EU directive on the subject. On the international level, in 2004 the Government hosted the international conference on renewable energy in Bonn (Renewables 2004). As to RES-E, the most important measures adopted were the 100.000 roofs programme for photovoltaics and above all the Renewable Energy Sources Act (RESA) adopted in 2000 and substantially amended in 2004.

The 100.000 roofs programme: Photovoltaic had not been able to develop much during the 1990s. The Red-Green Government wanted to provide new impulses. As the design of a new feed-in regulation was expected to take time, another market creation programme along the lines of the 100 MW wind and 1.000 roofs programme (both 1989) was adopted in January 1999 as a stopgap measure. It provided for reduced loans for PV roof installations, the goal being to achieve an installed capacity of about 300 MW. The programme was taken up slowly at first, but took off when RESA was introduced. By 2003, the two measures had led to installations of 350 MW. At that point, the 100.000 roofs programme was terminated and the PV market development turned over to improved feed-in tariffs.

Besides the Environmental Ministry, there is a second important actor, Ministry of Research and Education, which is currently implementing the strategy that includes also fostering environmental technologies[32].

3.3.4 Renewable Energy Sources Act of 2000

In April 2000, the Act on Granting Priority to Renewable Energy Sources was adopted in Germany, its declared purpose being to double RES-E production by 2010. This Act, which became one of the pivotal acts of the Red-Green coalition (Mez 2003), repealed the Feed-In Law of 1990 but maintained an essential feature, namely the reliance on feed-in tariffs to encourage the development of RES-E.

While under the Feed-In Law compensation rates were expressed as percentages of average end-customer tariffs, the new rates were now fixed for 20 years. For wind power, they were made dependent on the quality of the location: all operators would receive a favourable rate for at least five years and thereafter the rate would decline, but later in the case of less favourable locations. Rates were particularly favourable for PV, offshore wind and biomass. At the same time, there was now an annual decline in compensation for most sources, not for existing installations but for new installations and determined by the year they would go on line. A key regulatory element of the Act was the distribution of costs from RES-E compensation across all power grid operators on a pro rata basis, calculated on the ratio of RES-E in nationwide electricity sales. In addition, the utilities were now entitled to benefit from the special feed-in rates for their own RES-E generation facilities. This had not been the case earlier and might become lucrative for utilities, particularly in the case of highly capital-intensive investments such as those in offshore wind farms where they may beat back the new RES-E generators that arose in recent years.

3.3.5 RESA Amendment of 2004

After the re-election of the Red-Green coalition in autumn 2002, responsibility for RES changed from the Economic Affairs Ministry (held by a Social Democrat and always sceptical about RES-E) to the Environment Ministry (held by a Green); the Parliamentary Committee in charge changed in a parallel fashion. This opened new perspectives. The first draft by the Environment Ministry led to a lively conflict with Economic Affairs minister.

---

[32] http://www.hightech-strategie.de/en/36.php.

In the *Bundesrat*, the *Länder* ruled by Conservative governments opposed the Bill. The *Bundestag* (lower house) majority could simply have insisted on its earlier version. However, the Red-Green coalition negotiated with the Conservatives in an effort to secure support for maintaining RESA beyond 2007. Some of these wanted an expiration date of 2007 for the Act, or a declaration reversing the nuclear energy phase-out, while some criticised the 20% RES-E target for 2020. However the Conciliation Committee was content with more modest changes (exclusion of low-wind zones from the feed-in tariff), and the Bill was adopted in both houses.

Chief changes are a general strengthening of generators vis-à-vis the utilities; reduction of rates for onshore wind and exclusion of low-wind zones, but also improved rates for off-shore wind; inclusion of hydro plants up to 150 MW, and significant new incentives for biomass (especially small plants) with additional bonuses for innovative technologies (Bechberger and Reiche 2004). The most important factor was the increase in photovoltaics rates, which made them commercially attractive without additional support. This was introduced already in late 2003 and led to a veritable solar boom in 2004, expected to continue for several years.

3.3.6 Renewable Energy Sources Act of 2004

The new Renewable Energy Sources Act (Erneuerbare-Energien-Gesetz, EEG) of 21 July 2004 makes it obligatory for operators of power grids to give main concern to feeding electricity from renewable energies into the grid and to paying fixed prices for this. The approval of the precursor to the Renewable Energy Sources Act in 1990 triggered a major increase in wind power generation. The entry into force of the Renewable Energy Sources Act in 2000 led to a similar boom in biomass and photovoltaics. The use of geothermal energy for electricity generation has also developed significantly. The Renewable Energy Sources Act has thus proved to be an ideal and successful tool of energy policy[33].

4. Main Actors in the RES-E Arena

4.1 Germany

The Red-Green majority in power in 1998 was ambiguous on liberalisation at first. However, it came to accept this reform while still criticising the lack of regulatory oversight and the absence of sufficient environmental safeguards. Regarding the second point, some progress was made since the Red-Green coalition had a clear commitment to ecological modernisation of the energy sector. In

---

[33] http://www.erneuerbare-energien.de/inhalt/print/4306.

practice, however, policy changes were not comprehensive but rather incremental and restricted to a few key areas, e.g. nuclear, eco-tax, CHP and renewable energy policy. Each of these fields was managed by a different set of key government actors. These actors are far from presenting a uniform, polarised picture, as evident in the various decision-making processes.

In general, the parliamentary party groups are more reform-minded than some ministers, especially the minister of Economic Affairs, traditionally an advocate of the supra-regional utilities, of market liberalism and of industrial competitiveness. The political conditions changed in November 2005 and now a coalition of the Christian Democratic Party led by Chancellor Angela Merkel and the Social Democrats intends to continue the ecological modernisation. Its objectives are to increase the share of renewables in gross electricity consumption to at least 30% by 2020, followed by a continuous increase. Other goals are a 14% share of renewable energies in total heat supply by 2020 – more than double today's share – and a further increase in the share of biogenic fuels over the coming years. However, in the light of the current financial recession and accompanying developments, these objectives may be difficult to achieve.

4.2 Sweden

The two main bodies responsible for implementing energy policy measures are the Swedish Energy Agency and Affärsverket Svenska Kraftnät. However, the National Board of Housing, Building and Planning, the Swedish Consumer Agency, the Swedish National Electrical Safety Board, the Swedish Agency for Innovation Systems, the Swedish Research Council for Environment, Agricultural Sciences and Spatial Planning, the Swedish Research Council and the county administrative boards also help implement measures in the sphere of energy policy.[34]

4.3 Turkey

The Ministry of Energy and Natural Resources (MENR) is responsible for the preparation and implementation of energy policies, plans and programmes in co-ordination with its dependent and related institutions and other public and private entities. It reports directly to the Prime Minister.

The MENR has the following tasks and objectives:

- To determine and implement national energy policy objectives;
- to co-ordinate the dependent and related institutions and other public and private entities;

---

[34] http://www.sweden.gov.se/sb/d/5745/a/19594.

- to prepare and/or supervise programmes in conformity with the energy policy;
- to ensure the implementation of the programmes;
- to supervise and control all exploration, development, production and distribution activities for energy and natural resources.

The Research, Planning and Coordination Board (APK) of MENR co-ordinates the activities of the dependent and related institutions and executes national energy policy. It conducts long-term energy planning and develops different policy scenarios to support this work.

The General Directorate of Energy Affairs (EIGM) is the main policy-making body within the MENR. The EIGM is responsible for the co-ordination of the natural gas and electricity sector reform programmes, including the consequences of past efforts to bring private investments to the electricity sector. It also carries out studies on general energy and environmental policies, renewables and energy efficiency.

The General Directorate of Petroleum Affairs (PIGM) of MENR licenses oil exploration, production and refining. Since the abolition of the automatic pricing mechanism (APM) in the beginning of 2005, it no longer sets or controls oil prices.

The Electrical Power Resources Survey and Development Administration (EIE) of MENR is assigned to identify the energy potential of Turkish water resources and to prepare dam and hydropower plant projects. The EIE carries out various activities in relation to energy efficiency and renewable energy resources. The National Energy Conservation Centre (NECC) within the EIE is responsible for energy efficiency.

State Hydraulic Works (DSI) is the state water agency responsible for the development of all water resources in Turkey. DSI implements surface and ground water projects and plans, designs, constructs and operates dams and hydroelectric power plants for multi-purpose use.

The Turkish Atomic Energy Authority (TAEK) is the regulatory body responsible for the licensing of activities related to site selection, construction, operation and decommissioning of nuclear installations and other activities involving nuclear or radioactive materials. It also executes and supports nuclear R&D. The regulatory and R&D activities of TAEK will be separated in 2005 by creating an independent nuclear regulator.

The Energy Market Regulatory Agency (EMRA) was established as the independent regulatory authority for electricity by the Electricity Market Law in February 2001. After the enactment of the Natural Gas Market Law (May 2001) and the Petroleum Market Law (December 2003), EMRA was also given responsibilities in the natural gas and oil sectors. EMRA has administrative and financial autonomy and receives no funding from the state budget. Its total number of staff in September 2004 was 303, of whom 65 worked in the electricity department, 44 in the natural gas department, 32 in the petroleum department and 162 in other departments. EMRA collects its revenues principally from electricity and gas licensing fees and from a surcharge on electricity TPA tariff (maximum 1%).

The State Planning Organisation (DPT) is an advisory body of the Prime Minister. It assists the Government in determining economic and social objectives and the policies to be adopted. In practice, its major activities concerning the energy sector are the preparation of the five-year development plans together with the MENR and industry and preparing demand projections.[35]

5. References

Bilgen, Selçuk, Keleş, Sedat, Kaygusuz, Abdullah, Sarı, Ahmet and Kaygusuz, Kamil (2006). Global warming and renewable energy sources for sustainable development: A case study in Turkey.
http://ec.europa.eu/energy/res/legislation/share_res_eu_en.htm.
http://www.balkanlight.eu/abstracts_pdf/b13.pdf.
http://www.energy.eu/renewables/factsheets/2008_res_sheet_sweden_en.pdf.
http://www.erec.org/fileadmin/erec_docs/Projcet_Documents/RES2020/SWEDEN_RES_ Policy_Review_April_2008.pdf.
http://www.erec.org/fileadmin/erec_docs/Projcet_Documents/RES2020/SWEDEN_RES_ Policy_Review_April_2008.pdf.
http://www.erneuerbare-energien.de/inhalt/print/4306.
http://www.euractiv.com/en/energy/eu-renewable-energy-policy/article-117536.
http://www.europarl.europa.eu/meetdocs/2004_2009/documents/fd/d-tr20060425_06/d-tr 20060425_06en.pdf.
http://www.eva.ac.at/enercee/enlargement.htm.
http://www.hightech-strategie.de/en/36.php.
http://www.iea.org/textbase/pm/?mode=re&id=2547&action=detail.
http://www.invest.gov.tr/documents/publications/turkey2005.pdf.
http://www.planbleu.org/publications/atelier_energie/TR_National_Study_Final.pdf.
http://www.planbleu.org/publications/atelier_energie/TR_Summary.pdf.
http://www.rec.org/REEEP/energy_country_profiles/turkey.pdf.
http://www.regeringen.se/content/1/c6/06/47/22/2c000830.pdf.
http://www.regeringen.se/sb/d/5745/a/19594.

---

[35] http://www.invest.gov.tr/documents/publications/turkey2005.pdf.

http://www.sgu.se/sgu/eng/samhalle/energi-klimat/fornybar-energi_info_e.html.
http://www.sweden.gov.se/sb/d/5745/a/19594.
http://www.wealthdaily.com/articles/renewable-energy-germany/1418.
Jacobsson, S. and Lauber, V. (in press). The politics and policy of energy system transformation – explaining the German diffusion of renewable energy technology. *Energy / Policy*.

Chapter 3

Best Available Techniques – Working Paper – Biogas

Kerstin Kuchta, Konstantin Haker

The structure of this document is derived from an original reference document on Best Available Techniques (BAT). The determination of BAT is used as a tool to display the state-of-the-art of biogas technologies with the generally accepted guidelines of the IPCC Directive 2008(1/EC).

Scope

The scope of this document is the description of different techniques for the production of biogas that are successfully used in practice and which have proved value in business. In general this paper gives an overview of digestion techniques and shows briefly how the biogas can be treated to be used in different applications. The conversion of biomass into biofuels (e.g. ethanol, biodiesel, etc.) as well as production of landfill gas and sewage sludge gas will not further be discussed in this paper.

1. General Information

The production of biogas is based on the formation of methane that emerges during the biological degradation of organic matters. Processed organic materials (substrates) derive from agricultural and non-agricultural wastes (e.g. organics from municipal waste treatment and industrial waste treatment plants, and so called energy crops like maize, wheat and sugar beets. Products of the biogas process are biogas (mainly a mixture of methane and carbon dioxide) that has to be cleaned for further usage and a digestate which is usable as a fertilizer in agriculture. The cleaned biogas can be burnt in block heat and power plants where the combustion energy is transformed into heat, for district heating/fermentation tank heating and electricity that can be fed into the local grid. Cleaned biogas can also be fed into the local gas grid as a natural gas substitute without producing any electricity and heat directly on-site.

1.1 Biogas in Europe and Germany

Increasing prices for primary energy sources e.g. natural gas and oil are the reason why alternative, renewable energy sources are getting more and more important to the energy supply all over the world. Crucial for the increasing capacities of biogas production plants in the EU over the last years, are legal frameworks that offer economic incentives to operators, so that their engagement in energy production from biomass has positive effects in both economic and ecologic ways (Reinhold 2007).

Beside electricity and heat as a product of the anaerobic digestion (AD) process the emerging digestate can be utilised as fertilizer which leads to additional environmental benefits. The emerging digestate is due to its high content of nutrients an excellent fertilizer. By utilizing the digestate as a fertilizer a reduction in greenhouse gases is also attained since production of other fertilizers can be avoided. The methane emissions that manure normally causes can also be avoided. In a biogas plant nitrogen is converted to ammonium which is easier for plants to access than nitrogen bound in organic compounds. Therefore the leakage of nitrogen from the soil can be decreased which will increase the water quality in nearby waters. Digestion of manure also results in a decrease in odor and in a decrease in number of pathogens which also favors the utilization of digestate as a fertilizer compared to manure that has not been digested (Swedish Gas Centre 2009).

Since the 1990's roughly 110 biogas plants were installed in Germany. It were 3.500 plants at the beginning of 2007 with a total installed power of 1.100 MW. At the end of 2008 were 4.000 plants installed which cover 3% of Germany's natural gas demand 84 millard m^3 (Biogas e.V. 2009). For 2020 it is estimated to be 9.500 MW of total power installed in Germany (Reinhold 2007). This shows how rapidly development in the biogas sector in Germany proceeds (also see Figure 1).

This development is similar to the increase of biogas production in Europe. All over Europe the biogas production reached 6 Million tonnes in 2007 which is an increase of around 21% in respect to 2006. Targeted for 2010 is the production of 15 million tonnes of biogas. Estimations show that this target might not be achieved in 2010 (EurObserv 2008).

The German research centre on biomass in Leipzig estimates that one third of the natural gas demand in Europe could be covered by European biogas plants. This would be more than the amount of natural gas that Europe presently receives from Russia, so that biogas could be a full substitute for natural gas imported from Russia (Biogas e.V. 2009).

Figure 1: Number of agricultural biogas plants in Germany

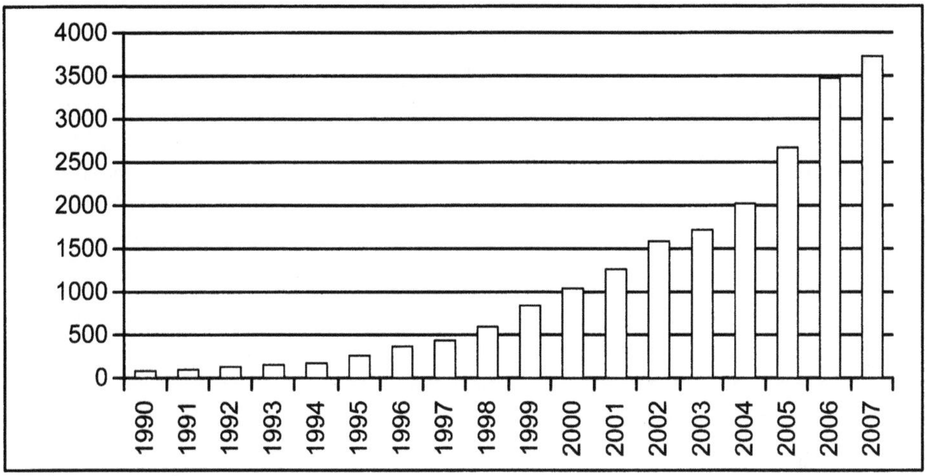

Source: Swedish Gas Centre 2009

Following Table 1 shows the development of biogas production from decentralised agricultural plants, municipal solid waste methanisation plants and centralised Combined Heat and Power (CHP) plants for 2006 and 2007 in ktoe per year.

Table 1: European biogas production 2006/2007

	2006	2007		2006	2007
Germany	1.011,7	1.696,5	Greece		
UK			Finland		
Italy	44,8	47,5	Ireland	1,8	1,7
France	3,6	3,7	Hungary	3,1	5,7
Spain	19,8	27,3	Portugal	9,2	15,4
Netherlands	47,1	82,8	Slovenia	0,4	3,8
Austria	103,4	126,4	Luxembourg	9,2	10,0
Denmark	62,0	62,6	Slovakia	0,4	0,5
Sweden	15,5	19,1	Latvia	0,0	0,0
Belgium	9,1	12,5	Estonia		
Czech Rep.	7,8	17,0	Lithuania	0,5	
Poland	0,5	0,5	Cyprus	0,0	0,2
Sum	1.325,3	2.095,9		24,6	37,3
Total EU	**1.349,9**	**2.133,2**			

Biogas production in: Decentralised agricultural plants, municipal solid waste methanisation plants, centralised CHP (Combined Heat and Power) plants. All values are given in ktoe.

Source: EurObserv 2008

1.2 Biogas in Turkey

Half of Turkey's energy demand is covered by natural gas and oil imports. Turkey's energy demand is currently covered mainly by fossil fuels, coals, oil and natural gas and less by geothermal and hydro power plants. There is no anaerobic digestion plant existent yet, respectively there is no information about operating digestion plants in Turkey available. But Turkey as a developing country in the field of anaerobic digestion has a high potential for the production of biogas, because of their relatively high organic content of their wastes. Also animal manure and some energy crops could be a good basis for an efficient production of biogas.

Turkey has a great potential of biomass and bio-energy production. Biomass energy seems to have a major potential for the usage as a energy source. The total recoverable bio-energy potential in Turkey was estimated to be around 16,92 Mtoe (million tonnes of oil equivalent), based on the recoverable energy potential from agricultural residues, livestock farming wastes, forestry and wood processing residues and municipal wastes. The primary energy consumption of Turkey is forecasted to reach 111,6 Mtoe in 2008. On the other hand, the share of renewable energy sources to primary consumption is estimated to be 1,5% in 2008. This share has not actually changed since 2006. Additionally, the contribution of energy production share of animal wastes and plant residues to primary energy consumption in Turkey ranged from 1,2% in 2006 to 1,0% in 2008 as well. In 2006, the share in electricity production from biogas to the total electricity generation was reported to be less than 0,1%, while for 2008, this share is forecasted to be 0,15%. It seems that, despite Turkey has a great potential of biomass to produce renewable energy and the law on utilization of renewable energy resources for the purpose of generating electrical energy has been brought into action in 2005 (Law No: 5346), the share of renewable energy in energy production is still low. Biogas production potential in Turkey was estimated to be around 1,5 to 2 Mtoe. However, since the share of renewable energy in energy production is so low, the possible contribution of biogas to this share can also be ignored. Actually, preliminary research activities using pilot-scale plants were initiated almost three decades ago by the General Directorate of Rural Services. These preliminary investigations covered production of biogas only from animal manure. However, these activities were somehow terminated in 1987. Besides, no research activity was encountered on production of biogas from agricultural residues and/or energy crops. By the way, there also exist not much data about applications of biogas technology in Turkey as well, in spite of a vast amount of research interest on biogas technology, especially in Europe (Demirel 2009).

2. Applied Techniques

In the following chapters an overview of technical equipment that is used for biogas production plants is given and according process conditions are presented in this paper. Different fermentation processes and their input materials as well as gas cleaning systems and gas utilisation opportunities will be furthermore discussed. Techniques considering the insertion of substrate, different gas storage systems, construction details on stirring units or engines that are able to utilise biogas are not part of this paper. Just basic applications and their characteristical mechanisms will be reckoned in the following.

2.1 Basics

The process of biogas formation (methanisation) can be generally divided into four main stages which are pictured in Figure 2: Different steps of biogas formation (FAL 2007).

Figure 2: Different steps of biogas formation

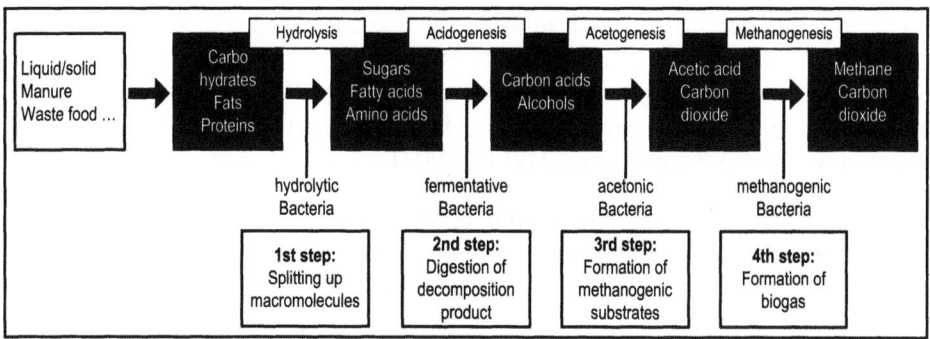

Source: FAL 2007

Liquid and/or solid substrates (e.g. liquid and solid manure, waste food, agricultural wastes, etc.), fats and proteins are hydrolysed in a first step. Emerging sugars fatty acids and amino acids are converted to carbon acids and alcohols in the acidogenesis phase. After this step, in the acetogenesis phase the carbon acids and alcohols are converted to acetic acids, hydrogen and carbon dioxide that in the last step, in the methanogenesis, are converted to mainly methane and carbon dioxide. The bacteria that are part of each conversion step are hydrolytic, fermentative, acetogenic and methanogenic bacteria (see also Table 2).

Table 2: Thermal stages and according aspired retention times

Thermal stage	Typical temperature	Aspired retention time
Psychrophilic	< 20°C	70 – 80 Days
Mesophilic	30 – 42°C	30 – 40 Days
Thermophilic	> 48°C	15 – 20 Days

Source: BayLfU 2007

The speed of methanisation depends on the temperature and the pH-value in the reactor. The temperature should always be above 20°C. Concerning a temperature beneath 20°C at so called psychrophilic conditions, substrates have a high retention time in the reactor (BayLfU 2007).

The pH-value should be between 6,1 and 8,1 which means that an acidification during hydrolysis (pH-values below 6,1) should be avoided, because the process then can completely break down. In biogas technology an operating temperature of 30°C-42°C (mesophilic conditions) is usual. In principal it is possible to reach higher degrees of degradation with a higher operating temperature (thermophilic conditions; temperature greater than 48°C), but this has as consequence that reactors would have to be build bigger and that the sensitivity of bacteria on changes in temperature is much higher so that small changes in temperature can lead to procedural problems or in the worst case to breakdowns, what both would lead to much higher process costs (BayLfU 2007). The dependence on process temperature and relating biogas yields is shown in Figure 3: Biogas production relating to temperature (BayLfU 2007).

Figure 3: Biogas production relating to temperature

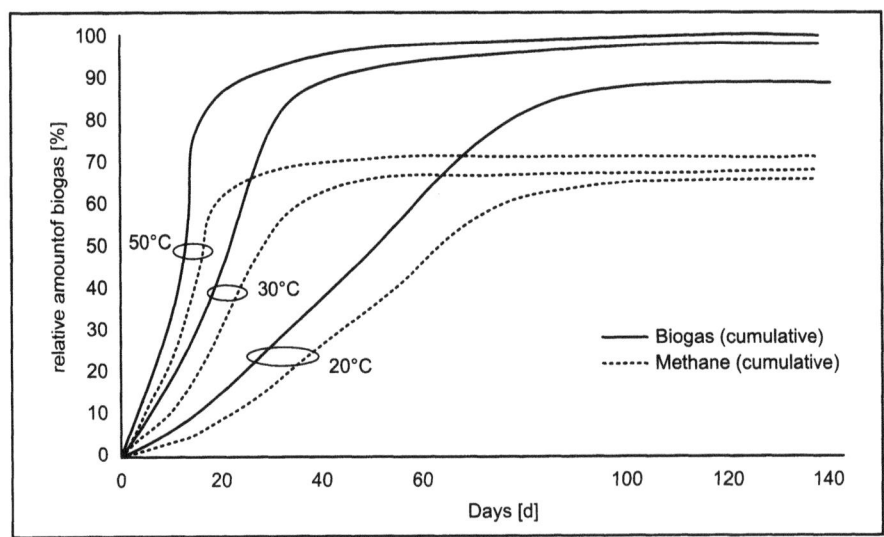

Source: BayLfU 2007

During the methanogenesis occur other reactions that influence the gas quality in the output, because not only methane and carbon dioxide are produced. Depending on the input material hydrogen sulphide (H_2S) and ammoniac (NH_3) can be built which are supposed to be harmful to human health and to lead to damages inside the engine of the block heat and power plant (BayLfU 2007). From step 1 (Hydrolysis) to step 4 (Methanogenesis) intermediate catabolic products of previous steps are feed material for the following step. E.g. product of Hydrolysis are short-chain peptides, long-chain fatty acids, glycerine, amino acids and monosaccharide's which are feed material for acidogenesis and so on and so forth.

Table 3 which consists of four phases of biogas formation, substrates and bacteria (BayLfU 2007) gives an overview of the four process phases, their different input substrates with examples of bacteria that are active and the resulting products of each step.

Table 3: Four phases of biogas formation, substrates and bacteria

Phase	Substrate	Examples of Microorganisms	Products
Hydrolysis	Carbon hydrates Proteins Fats	Clostridium spp. Bacillus spp. Pseudomonas spp.	Short-cahin peptides Long-chain fatty acids Glycerin Amno acids Monosacharides
Acidogenesis	Short-cahin peptides Long-chain fatty acids Glycerin Amino acids Monosacharides	Clostridium spp. Bacteroides spp. Butyrivibrio spp.	Volatile fatty acids (acetate, Probionate, butyrate) Aldehydes, alcohols Ketones, ammoniac Carbon dioxide, hydrogen
Acetogenesis	Volatile fatty acids (acetate, Propionate, butyrate) Aldehydes, alcohols Ketones, ammoniac Carbon dioxide, Hydrogen	Clostridium spp. Eubacterium spp.	Acetate Carbon dioxide Hydrogen
Methanogenesis	Acetate Carbon dioxide Hydrogen	Methanosarcina spp. Methanosaeta spp. Methanobacterium spp.	Methane Carbon dioxide

Source: BayLfU 2007

2.2 Basic information on biogas

In biogas plants different substrates of differing origin are being used. Accordingly different techniques for the treatment of substrates and diverse techniques for substrate insertion and different types of fermenters and operating procedures can be found in biogas production sector. Besides, depending on the scale of plants different techniques for gas processing, gas storage and gas utilization are applied.

Biogas appliances can roughly be divided in liquid and solid substrate fermentation (LSF/SSF) or also called wet and dry fermentation processes, what could lead to misunderstanding, because in reality, anaerobic digestion can only happen in the liquid phase. Limit for the diversion of LSF and SSF is the content of solid material. Roughly from 16% solid substrate content on, the process is said to be a SSF process (BayLfU 2007). Consecutively the terms LSF and SSF are used.

Figure 4: Substrates for a biogas plant

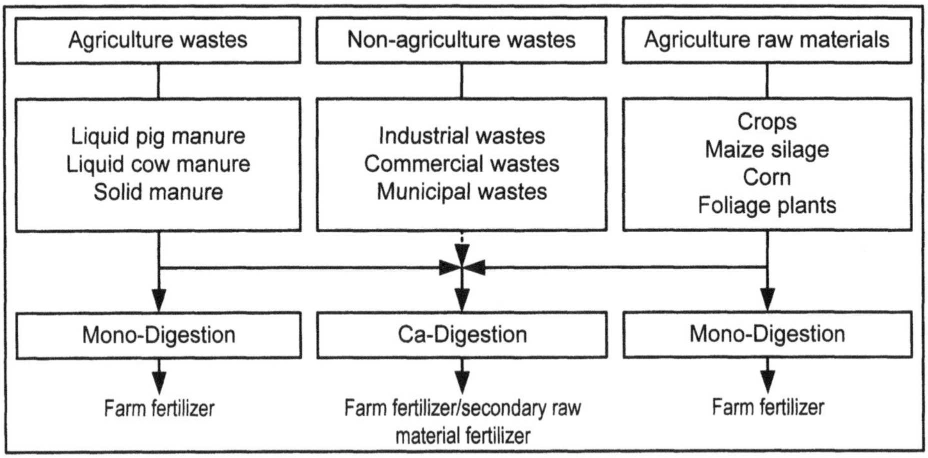

Agriculture wastes	Non-agriculture wastes	Agriculture raw materials
Liquid pig manure Liquid cow manure Solid manure	Industrial wastes Commercial wastes Municipal wastes	Crops Maize silage Corn Foliage plants
↓	↓	↓
Mono-Digestion	Ca-Digestion	Mono-Digestion
↓	↓	↓
Farm fertilizer	Farm fertilizer/secondary raw material fertilizer	Farm fertilizer

Source: FAL 2007

Biogas emerges by degradation of organic matter by anaerobic digestion. Figure 4: Substrates for biogas plant (FAL 2007), shows different possible input materials for a biogas production plant. A co-digestion of organic matters with different origin is also being applied as well as mono-digestion appliances for agricultural waste and agricultural raw materials.

Emerging biogas can then be further treated either to be fed into the gas grid or for the use as a fuel for vehicles with gas engines or for the direct combustion in block heat and power plants. The reason for the necessity of further gas treatment systems is the varying content of each gas components, that depend on the fermented substrate. Methane and carbon dioxide as well as contents of ammoniac and sulphur in form of hydrogen sulphide (H_2S) emerge as products of the anaerobic digestion process, that can lead to damages during the combustion process and/or are harmful to human health (RP 2009).

Table 4: Typical composition of biogas

Component	Percentage	Characteristics of Biogas
Methane	40-75	– Density: 1,2 kg/m³ – ignition temperature: 700°C – ignition concentration: gas content 6-12% – odour: smells like foul eggs, because of H_2S content – calorific value: 14,4-27 MJ/kg (depending on methane content)
Carbon dioxide	25-55	
Water vapor	0-10	
Nitrogen	0-5	
Oxygen	0-2	
Hydrogen	0-1	
Hydrogen Sulphide (H_2S)	0,002-2	
Ammoniac	0-1	

Source: RP 2009

2.3 Common techniques applied in the sector of biogas production

As aforementioned the procedure of biogas production consists of different procedural steps and the applied machine and equipment technology shows a high variety. Inside this variety exist many different opportunities to combine these single technology components. This is the reason why during the planning phase of an installation of a biogas plant experts should be involved concerning the choice of fitting appliances and the suitability of the technical equipment. The adjustment between technical equipment, utilised substrates and operation management of the plant are determining the gas quality in the output. And at last decides the substrate about the usage of a specific technique and their design (dimensioning, reactor tanks, etc.), e.g. pasteurization, crushing, dimensioning of pipes, pumps, gas processing, gas storage, block heat and power plant, etc. (BayLfU 2007).

2.3.1 Material acceptance and registration

The step of material acceptance and registration is necessary to plants that beside their own organic materials digestate external organic matters. In that case the registration and the pre-treatment of organic wastes is essential for bio-waste processing. Registration and pre-treatment of incoming material include following steps:

- Classification and registration of delivered bio wastes;
- separation and removal of contraries;
- interim storage in buffer stock;

- substrate conditioning (e.g. addition of water, liquid co-enzymes or manure to the substrate to get pumpable material);
- if necessary, sanitation of substrates at minimum 70°C with a retention time of minimum 1 hour;
- if necessary hydrolysis and pre-acidification in the hydrolysis tank.

Classification and registration as well as interim storage are necessary to mix the bio-waste in such a way that a continuous operating mode of the plant can be ensured. Hereby the size of the stock should depend on delivery intervals and relative frequencies. Besides, storage management must be anticipated in a way so that the mixing of hygienic critical and hygienic harmless materials is avoided. If smelly wastes like for example waste grease is being stored, storage areas should be roofed and if necessary constructed with an exhaust air cleaning system, to reduce odor emissions to a minimum. Separation and removal of contraries is to guarantee the functionality of the process and to avoid problems in handling the biomass (BayLfU 2007).

2.3.2 Substrate conditioning

To achieve a continuously high biogas production the infeed of the fermenter tank is of highest importance. This means, the substrate composition should not be exposed to too high variations, because microbiology has to readjust to each new substrate composition what could lead to losses in yield of biogas. An additional aspect brings the conservation of renewable primary products. Because all year long renewable primary products are not available in case of co-digestion this material has to be conserved for times where supply is not possible so that these can be fed from stock into the fermenter. Energetically, best practice is the ensilage of renewable primary products.

Crushing units are usually applied to increase the surface of the material to give bacteria a bigger working surface so that degradation processes are accelerated and retention time of the substrates can be shortened. This is especially important to long fibrous substrates and has as an additional benefit that problems such as reduction of pump ability, congestion of pipes, fixation of the agitators and increasing formation of layers of scum can be avoided (BayLfU 2007).

Pasteurization is necessary for biogas plants that are operated with bio wastes that are hygienic dubious. Directives concerning the need for sanitation are given by the European Directive No. 1774/2002 and regards animal byproducts that are not convenient for human consumption. To decide whether the necessity for pasteurization is present please consult according European and/or particular local or regional legislation. If a pasteurization is required the material is usually pasteurized before fermentation. The material is pasteurized at 133°C and 3 bar for 20 minutes (sterilization by compression). Other byproducts, after crushing to a particle size of minimum 12 mm, can be sterilized at 70°C within 60 minutes retention time. During pasteurization temperature, fill level, pressure and retention time are monitored and documented to guarantee documentation of the processes without a gap. In general pasteurization is applied in two different procedures. A quasi-continuous and a continuous pasteurization process (BayLfU 2007). In this paper pasteurization will not be mentioned more detailed, because the focus here lies in biogas production procedures and not in the optimal conditioning of the substrate.

2.3.3 Fermenter technique and construction

Fermenter technique regards to the setup of a reaction tank. This means part of the fermenter technique are the feed systems (infeed technique, pumps, pipes, dense medium and fermenter discharge), heating systems, homogenization system and, process measuring and control systems. Besides, types of fermenter constructions are divided into the kind of infeed (batch, storage, exchangeable containers, flow and flow-storage), the number of process steps, temperature and the dry matter content of the substrate (see also Table 5).

Table 5: Types of infeed procedures and their options

Infeed	Number of process steps	Temperature	Dry matter content of substrate
Batch, storage, exchangeable tanks, flow and flow-storage	One, two or more	Psychrophilic, mesophilic, thermophilic	Wet and dry fermenters

Source: BayLfU 2007

In biogas sector usually single-stage processes are applied. Two or multi-stage processes are commonly applied in industrial respectively in waste management plants (BayLfU 2007).

Following table gives a short overview of procedural steps of each infeed procedure. The pictures show a simplified schema of each procedure and its main characteristics.

Table 6: Different feeding procedures

Infeed procedure	Application flow	Comments
Batch	VG F.1 Lb — Emptying phase; VG F.1 Lb — Filling phase; VG F.1 Lb — Fermentation. Vg: Slurry store, F.x: Fementer x, Lb: Storage Tank	No continuous biogas production, safe retention time, no short circuit fluid stream, low digester efficiency.
Storage	VG F.1 Start — or — VG F.1 End	Fermenter here is the same as storage tank., no continuous gas production, high retention time, medium digester efficiency, high tank volume.
Exchange-able Tanks	VG F.1 F.2 Lb Fermentation Tank 2; VG F.1 F.2 Lb Fermentation Tank 1; VG F.1 F.2 Lb Fermentation Tank 1	Superposition of two settling fed fermenters, more continuous production of biogas than in batch operation, safe retention time, low digester efficiency.
Flow	VG F.1 Lb	Continuous flow of substrate, constant biogas production, no safe retention time, high digester efficiency.
Flow-storage	VG F.1 Lb	Fermenter in continuous flow, constant biogas production, no safe retention time, high digester efficiency.

Source: BayLfU 2007

Usually there are vertical and horizontal fermenters produced that have different advantages and disadvantages. Vertical fermenters usually have a round cross-section and are either made of stainless steel (volume from 100m³ to 3.500 m³) or in case of cost reduction and for better statics made of concrete (volume ranges from 100 m³ to 5.000 m³). Vertical fermenters are completely mixed so that a contamination of fresh fed substrate is assured. Additional actions to inoculate fresh fed substrate are not necessary (BayLfU 2007).

2.3.4 Fermenter and feeding techniques

Feeding technique

To feed the biogas reactor or fermenter different technologies are applied to move the substrate into the reactor. Figure 5 pictures three different feeding techniques.

Figure 5: Different feeding mechanics

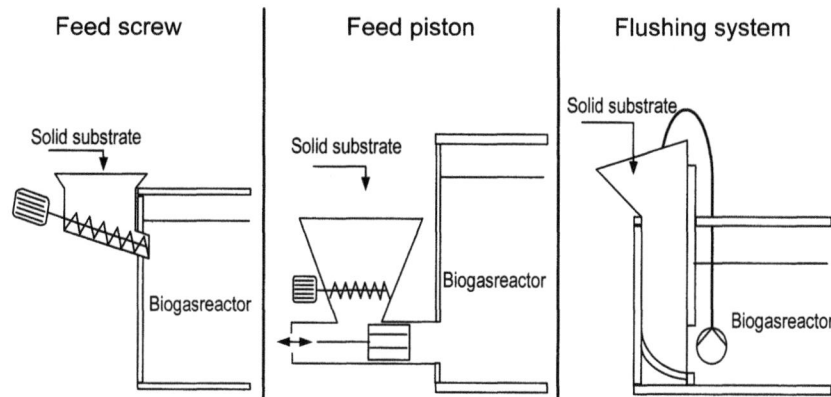

Source: FAL 2003

Fermenter design

Basically are two different types of fermenters, horizontal and vertical tanks, applied in biogas production sector. Horizontal fermenters usually have a rectangular or quadratic cross section. Typically they are built in a size from 250 m³ to 1.000 m³ tank volume. Because of their length In comparison to vertical fermenters, horizontal fermenters have higher costs concerning construction and insulation, but the low height of the tanks allows it to install slow turning stirring units, that work across the direction of flow and effects an optimal mixing inside the reactor up to a solid substrate content of 20%. The setup of a typical fermenter in vertical design is shown in Figure 6. There are many different opportunities for the installation of the stirring unit and depends on the kind of substrate that should be treated.

Figure 6: Fermenter in vertical design made of concrete

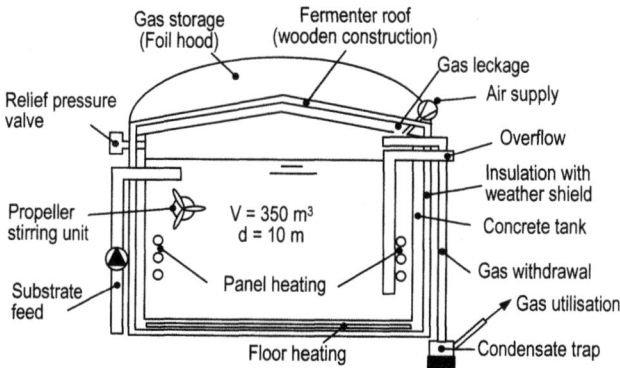

Source: BayLfU 2007

Horizontal fermenters where the content is only transported by the addition of fresh substrate via the crowding-out effect are called plug-flow reactors. The substrate is moved through the tank like a cork what effects areal isolated degradation zones inside the tank and avoids the occurrence of short circuit flows. This results in an increase of substrates sanitation ability and has positive effects on the grade of degradation. Missing inoculation of the fresh substrate that comes not into contact with bacteria rich material inside the fermenter can be compensated by the addition of liquid and/or solid manure or by refeeding the substrate.

Figure 7: Horizontal concrete fermenter

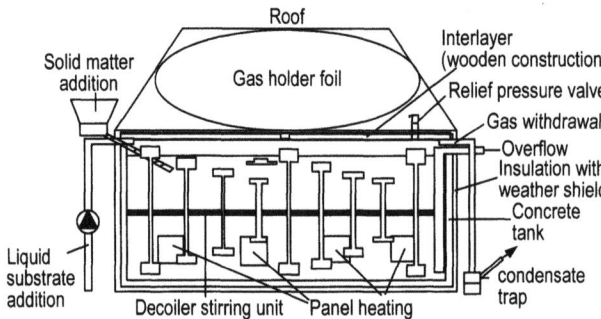

Source: BayLfU 2007

Besides, special types of biogas production systems e.g. the garage system or the cavern system exist. Both are applied for solid matter fermentation in discontinuous operating mode.

Figure 8: Setup of a garage system

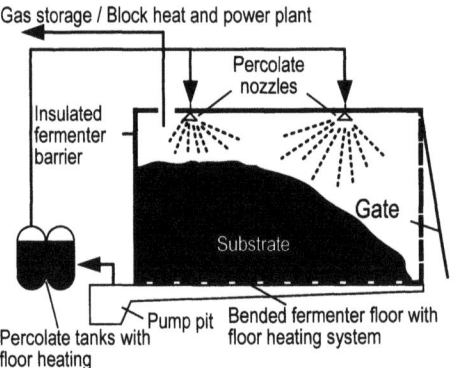

Source: BayLfU 2007

Discontinuous fermentation of solid matter. Volume per garage (= fermenter) 100 m³. Only 2/3 of the volume can be exploited because of the feeding technique with a wheel loader. The gate is gas proof and lockable. To supply the substrate with anaerobe bacteria sufficiently and to buffer emerging acids it is mixed with fermented substrate with a ratio of 40% (fermented substrate) up to-60% (fresh substrate).

Besides, exist two other procedures in biogas production, the mine system and the film tube system. Because these systems are relatively uncommon they are described more detailed in the following.

2.4 Gas cleaning

An important procedural step makes the gas cleaning system. As aforementioned, beside methane (CH_4) and carbon dioxide (CO_2) biogas also contains traces of sulfur dioxide (H_2S). This gas harms humans and the environment, (at concentrations from 150 ppm on olfaction fails, from 250 ppm on pulmonary oedema can accumulate and from concentrations of 1.000 ppm on H_2S acts deathly in shortest time) and causes damages in the process equipment, because H_2S acts highly corrosive on building and technical equipment. The amount on H_2S in the biogas only depends on the used substrate and usually lies 0,02 and 0,5 Vol.%, but in extreme cases contents of 1,5% (1.500 ppm) were measured, so that its content in biogas has to be reduced and a gas cleaning system is essential.

Procedures for desulfurization are divided in three main categories: physicochemical, biological and combined methods. Physicochemical procedures are precipitation, absorption, adsorption, oxidation, pressure cleaning and disconnection by membranes. Biological methods cover removal directly inside the reactor, bio-scrubber and bio-filters. As the combined method is known lye scrubbing in connection with biological oxidation (BayLfU 2007).

3. Emissions and consumptions

Air pollutants from biogas plants mostly emerge from the combustion of the biogas in block heat and power plants whereas as much as possible of the generated heat should be used. Alternatively biogas can be utilised in boiler plants or in gas turbines. After further gas cleaning the gas can be fed to the regional gas grid and be used like natural gas. Typically carbon oxides (CO_x) and nitrogen oxides (NO_x), sulphur dioxide (SO_2) and some unburnt hydrocarbons are emitted after the combustion of biogas in combustion engines. If pilot injection engines are being used also relevant dust emissions (soot) must be anticipated (BMWA 2007). Depending on kind of utilised substrates and process and plant planning also odour emissions occur. These emissions can be kept small by covering fermenters, conveying systems and storage facilities.

The biogas process needs electricity e.g. for feeding systems and stirring units and heat to warm the reactor to reach process temperature depending on the kind of process. Detailed consumption data cannot be given, because the amount of energy consumption depends on size of the plant and the installed technical equipment.

4. Techniques to consider in the determination of Best Available Techniques

Additional techniques are considered generally to have potential for achieving a high level of environmental protection in the biogas production sector. For example management systems, process-integrated techniques for process measurement (e.g. gas analysing systems, pH-control systems in the reactors and gas utilization systems), have significant influence on plants efficiency and therefore on the environment. Prevention, plant control, design, management and gas utilization procedures as well as material collection systems and efficient energy usage must be considered to achieve the objectives of IPPC. Because it is not possible to be exhaustive and because of the dynamic nature of biogas industry, it is possible that there may be additional techniques left unmentioned, but which may also be considered as Best Available Techniques (BAT).

5. Best Available Techniques

Best Available Techniques in biogas sector depends on the targeted product and the required product quality, the type and appearance of the input material and economic factors like market situation for specific materials (e.g. prices for metals, glass, paper and wood.)

Depending on the kind of biogas plant BAT is use aforementioned techniques (see 2 Applied Techniques) in such way that the gas production rate reaches the maximum and that emissions to air, water and soil are minimal. It is not possible to define BAT for one special kind of biogas plant, because all the techniques depend on bases of operations as there are kind of organic wastes or substrates that are used, the location of the plant, size of the plant, neighbourhood, energy demand and necessary energy output, etc. Only a well thought combination of different feeding, fermentation, gas storage and utilisation techniques meets the optimal degree of efficiency and then leads to an environmental and economic benefit.

Management systems like environmental management systems, raw material and utility management systems, knowledge of waste IN and waste OUT and therefore management of waste analysis are also considered to be BAT for a biogas plant. For further information concerning BAT in waste treatment methods please consult the BAT for waste treatment installations.

6. Emerging Techniques

Highest potential for emerging techniques lies in the development of gas cleaning systems and the improvement in measuring and process control techniques. An important role in research also plays the usability of biogas in different appliances. As an example the usage of biogas in fuel cells is one of many research fields. A detailed and exhaustive collection of emerging techniques in the biogas sector is not possible yet, but the german research centre on biogas is working on a BAT paper for biogas plants that should be a basis for the so called Sevilla process which has the goal to edit an IPPC directive that reaches validity all over Europe. At this time there is a draft with date 15 October 2008 (only in german language) available on the internet (DBFZ 2008).

7. Glossary

Terms and definitions

All definitions were taken from (RP 2009):

Biogas is a mixture of methane, carbon dioxide and a small amount of rest gas. To get further information of the composition of biogas see Table 4 in chapter 2.

A *Biogas plant* is a plant where organic material is fermented and biogas is produced, collected, stored and recovered (energetically). All different technical equipment that is necessary for operating the plant are part of a biogas plant.

The *fermenter* (also *fermentation tank*, *digestion tank* or *reactor*) is a gas proof tank where microbiologic degradation of the substrate takes place. It is usually equipped with a heating system and a mechanical stirring system and is encircled by a thermal insulation layer. Optionally the fermenter is connected upstream to a *dump* where the material can be stored since it is processed to the fermenter.

Independent from the dump it is possible to utilise a *dosing feeder for solid material* to add (dry) material to the fermenter via screws or extrusion dies.

Fermentation is a stepwise enzymatic degradation of organic matter under anaerobic conditions (without presence of oxygen). In wet environment digestion gas is generated. Different bacteria crack organic substances (hydrolysis), produce organic fatty acids (acidification) and convert these into methane (methanisation).

Substrate is the input material which is fermented. It consists of organic materials, e.g. solid and liquid manure) or especially for the use in biogas plant produced renewable raw materials like so called energy crops (maize, wheat, etc.) or grass silage and co-ferments.

Co-ferments (also *co-substrates*) are organic matters that are not generated by agriculture, e.g. kitchen wastes, contents of fat separators and different residues from agro-industry such as pomace, draff, sugar beet pulp, blackstrap molasses and oil seed residues.

Co-fermentation is the fermentation of wastes from agriculture and wastes that do not come from agriculture. In this way gas yield can be increased significantly.

In a *block heat and power plant* mechanical energy and heat are produced by the combustion of biogas in a combustion motor. Heat is being used as thermal heat and the mechanical energy is converted into electricity via a generator.

The *gas storage* is a gas proof tank or a foil bag where the biogas is temporarily stored.

Manure storages are opened, covered or closed tanks or basins where manure, slurry and the fermented substrate is being stored.

8. References

(BayLfU 2007)	Biogashandbuch Bayern, edited by Landesamt für Umwelt und Naturschutz Bayern, Germany, 15th July 2007.
(Biogas e.V. 2009)	Fachverband Biogas e.V. (German biogas Association), 16.01.2009, www.biogas.org/datenbank/file/notmember/presse/09-01-16_PM_IGW.pdf, last visit 16th February 2009.
(BMWA 2007)	German federal ministry for economy and work, "Technical basics for the assessment of biogas plants", 2007.
(DBFZ 2008)	Deutsche BiomasseForschungsZentrum gemeinnützig GmbH, "Stand der Biogastechnik", Entwurf vom 15.10.2008.
(Demirel 2009)	Demirel, Burak: "The potential and opportunities of biogas use in Turkey", Bogazici University, Institute of Environmental Sciences, February 2009.
(FAL 2003)	Weiland, P. / Rieger, Ch. / Ehrmann, Th.: „Evaluation of the newest biogas plants in Germany with respect to renewable energy production, greenhouse gas reduction and nutrient management", Institute of Technology and Biosystems Engineering Federal Agricultural Research Centre (FAL), Future of Biogas in Europe II, Esbjerg 2-4 October 2003.

(FAL 2007)	Bundesforschungsanstalt für Landwirtschaft (Federal Research Institute for agriculture): Forum bioenergy villages, Göttingen, March 2007.
(Gazio 2007)	N. Kizilaslan, H. Kizilaslan: Department of Agricultural Economics, Faculty of Agriculture, Gaziosmanpasa University, Turkey; "Energy Sources", Part B: Economics, Planning, and Policy, Volume 2, Issue 3 July 2007, pp. 277-286.
(Reinhold 2007)	Reinhold, Günther: „Biogas – Technik und Trends", Thüringer Landesanstalt für Landwirtschaft, Pressemitteilung Nr. 40, September 2007.
(RP 2009)	Manual on the construction and operation of biogas plant, ministry of environment and forest, german federal state Rheinland-Pfalz, http://www.mufv.rlp.de/ fileadmin/ img/ inhalte/ luft/Biogasanlagen_Handbuch.pdf, last visit: 24 February 2009.
(Swedish Gas Centre 2009)	Petersson, Anneli: "Biogas from international perspective", Swedish Gas Centre, http://www.sgc.se/rapporter/resources/Biogas_International_Perspective.pdf, last visit 17 February 2009.

Chapter 4

Best Available Techniques – Working Paper – Biofuels

Kerstin Kuchta, Konstantin Haker, Georgi Chobankov

The structure of this document is derived from an original reference document on Best Available Techniques (BAT). The determination of BAT is used as a tool to display the state-of-the-art of biofuels technologies with the generally accepted guidelines of the IPCC Directive 2008(1/EC).

Scope

The scope of this document is the overview and the brief description of different techniques for production and use of biofuels that are successfully applied in practice. In general this paper gives an overview of the first and second generation of biofuels as well as examples of applications of solid biofuels in praxis.

1. General Information

Usually biofuels are divided in two different generations. First and second generation of biofuels are explained in the following:

1.1 First Generation Biofuels

First generation biofuels are biofuels which are on the market in considerable amounts today. Typically first generation biofuels are sugarcane ethanol, starch-based or 'corn' ethanol, biodiesel and Pure Plant Oil (PPO). The feedstock for first generation biofuels production either consists of sugar, starch and oil bearing crops or animal fats that in most cases can also be used as food. A first generation biofuel is characterized either by its ability to be blended with petroleum-based fuels, combusted in existing internal combustion engines, and distributed through existing infrastructure, or by the use in existing alternative vehicle technology like FFVs ("Flexible Fuel Vehicle") or natural gas vehicles. The production of first eneration biofuels is commercial today, with almost 50 billion litres produced annually worldwide. There are also other niches for biofuels, such as biogas which have been derived by anaerobic treatment of manure and other biomass materials. However, the volumes of biogas used for transportation are relatively small today (IEA Bioenergy 2008).

1.2 Second Generation Biofuels

Second generation biofuels are those biofuels produced from cellulose, hemicellulose or lignin. A second generation biofuel can either be blended with petroleum-based fuel, combusted in existing internal combustion engines and distributed through existing infrastructure or is dedicated for the use in slightly adapted vehicles with internal combustion engines e.g. vehicles for DME (Di-Methyl Ether Vehicle) Examples of second generation biofuels are cellulosic ethanol and Fischer-Tropsch fuels (IEA Bioenergy 2008).

The future of biofuel which is made from biomass is of very high interest worldwide. Global biofuel production has tripled from 18,7 billion litres in 2000 to about 60,6 billion in 2007, but still accounts for less than 3% of the global transportation fuel supply. About 90% of production is concentrated in the United States, Brazil, and the European Union (EU). Production could become more dispersed if development programs in other countries are successful. The leading raw materials, or feedstocks, for producing biofuels are corn, sugar, and vegetable oils (AmberWaves November 2007).

Figure 1: Global biofuel production between 2000 and 2007

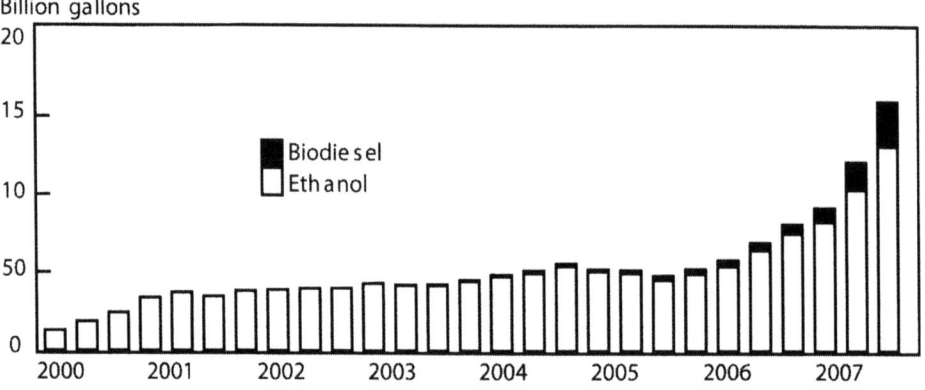

Source: International Energy Agency; FO Licht

While rapid expansion in biofuel production has raised expectations about potential substitutes for oil-based fuels, there have been growing concerns about the impact of rising commodity prices on the global food system. According to the International Monetary Fund, world food prices rose 10% in 2006 because of increases in corn, wheat, and soybean prices, primarily from demand-side factors, including rising biofuel demand. The Chinese Government put a moratorium on expanded use of corn for ethanol because of rising feed prices and is promoting other feedstocks that do not compete directly with food crops, such as cassava, sweet sorghum, and jatropha (an oil-bearing plant originally from South America) (AmberWaves November 2007).

Figure 2: Worldwide Biofuels Production (FO Licht)

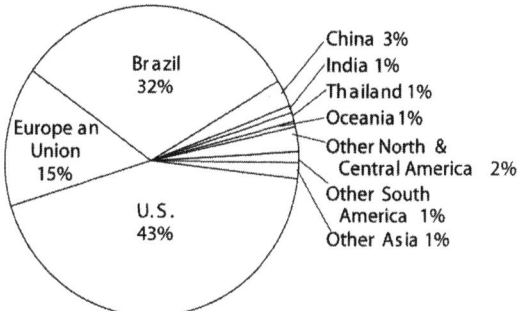

Source: FO Licht, includes only ethanol for fuel.

1.3 Biofuels in Europe and in Germany

Increasing prices for primary energy sources e.g. natural gas and oil are the reason why alternative, renewable energy sources are getting more and more important to the energy supply all over the world. Crucial for the increasing capacities of biofuels production plants in the EU over the last years, are legal frameworks that offer economic incentives to operators, so that their engagement in energy production from biomass has positive effects in both economic and ecologic ways. More than four years after its implementation, the European directive on promotion of biofuels intended for transportation purposes has made it possible to reach consumption of approximately 8,1 M toe in 2007. This figure now represents 2,7% of the energy content of all of the fuels used in road transport in the European Union. Biofuels consumption continues to increase in the European Union, but at a less buoyant rate than in 2006. It grew by 44,5% between 2006 and 2007 (+2.504.917 toe). The growth rate was 79,7% between 2005 and 2006 (+ 2.494.221 toe). It should be specified that the figures for 2006 and 2007 have been partially consolidated with the help of the reports that each State must transmit to the EU in order to verify progress made in terms of the directive. Looked at this in greater detail, it can be seen that biofuel consumption has risen from 4.083.191 toe in 2006 to 6.091.535 toe in 2007. Production of bioethanol, whether it is mixed with petrol or transformed beforehand into ETBE (product composed of 47% bioethanol), has increased from 880.443 to 1.225.668 toe

(+ 39,2%). The other types of biofuels, essentially represented by vegetal oil, have grown by 19.7%, going from 661.587 to 791.935 toe (EurObserv 2008).

Germany is European leader for consumption of biofuels used for transport. It consumes nearly 4 million toe, i.e. 49% of EUs' total consumption. More than 750.000 toe of vegetal oil (838.000 tons) and nearly 300.000 toe of bioethanol (460.000 tons) is added to round about 3 million toes of biodiesel (3.318.000 tons) consumed in Germany which had already passed the European directive objectives as early as 2006. In 2007, biofuels represented 7,3% of total consumption of fuel used in the transport sector. The German law on biofuels (Biokraftstoffquotengesetz), which entered into application on 1 January 2007, put an end to the total exemption of taxes on biofuels. It still obliges fuel distributors to integrate a minimum percentage of biofuels in classical fuels. This amount is established at 4,4% of biodiesel in diesel oil and in 2007, at 1,2% of bioethanol in petrol. This last percentage is estimated to rise to 2% in 2008, to 2,8% in 2009 and to 3,6% in 2010. Beginning in 2009, a combined supplementary quota of 6,25% will be applied for the two fuels. It will then gradually rise to 8% in 2015. The law also provides a degressive tax incentive for second generation biofuels, for fuel biogas and for E85 (85% ethanol, 15% gasoline) up to 2015 (EurObserv 2008).

1.4 Biofuels in Turkey

In anticipation of these growing demands in the EU as well as its own entry into the EU, Turkey has modified its laws to permit sale and use of biofuels. Today a blend of 2% biodiesel in regular diesel is permitted, compared to EU law which will soon permit up to 10%. Various companies in Turkey have begun active Research and Develpment programs to achieve compliance with the higher EU targets. Development of the knowledge base and production capacity in Turkey for biodiesel production represents a huge opportunity, both for Turkish domestic consumption as well as for technology and product exports to the EU and elsewhere. The key drivers ultimately promoting development of the biofuels market are the same for energy in general: concerns regarding security of supply (biofuels allow a more diversified mix, even if imported), climate change (carbon emissions are less than burning fossil fuels), and costs (although reliant on subsidies currently, the expectation is that with carbon taxes eventually factored in, biofuels will be cost competitive). The tax and subsidy regimes in different countries will in meantime continue to play a pivotal role in the development of biofuels industry. The data and the research findings about biodiesel and its potential for development in Turkey vary somewhat across different organizations, academics and Governmental entities who have worked on that issue. The main reasons for these variations are:

- Rapidly increasing interest in biodiesel;
- largely unregistered and unreported biodiesel production, estimated indirectly from other data, such as oil imports;
- ongoing development of government policies and procedures to properly regulate this industry.

Vegetable oil, hence oil seed, is the main raw material used for biodiesel production. Turkey is a net importer of both of these materials. According to the Turkish Ministry of Agriculture, in 2004 oil seed was planted in 7% of the total cultivated land. The government is aiming to double this percentage. Turkey's oil production at the end of 2004 was approximately 500.000 tons while consumption was 1.300.000 tons, leaving a gap of 800.000 tons which was met by imports. By 2006, imports had increased to 954.000 (The Wharton School of the University of Pennsylvania 2007).

2. Applied Techniques

In the following successfully applied techniques in industry are shown and briefly explained. Only short descriptions of the different processes can be given here, because each production plant has its special setup and target fuels with its certain by-products. Focus lies on the production processes of biodiesel, bioethanol, the conversion of biomass and applications in syngas production and application.

2.1 Basics

2.1.1 Biodiesel

There are three basic ways for biodiesel production from oils and fats:

- Base catalyzed transesterification of the oil;
- direct acid catalyzed transesterification of the oil;
- conversion of the oil to its fatty acids and then to biodiesel.

Catalyzed transesterification is mainly realised in biodiesel production sector. The reasons therefore are the following:

- It is a low temperature and pressure process;
- it yields high conversion (98%) with minimal side reactions and reaction time;
- it is a direct conversion to biodiesel with no intermediate compounds.

The chemical reaction for base catalyzed biodiesel production is depicted below (Figure 3). One hundred pounds of fat or oil (such as soybean oil) react with 10 pounds of a short chain alcohol in the presence of a catalyst to produce 10 pounds of glycerin and 100 pounds of biodiesel. The short chain alcohol, signified by ROH (usually methanol, but sometimes ethanol) is charged in excess to assist in quick conversion. The catalyst is usually sodium or potassium hydroxide that has already been mixed with the methanol. R', R", and R"' indicate the fatty acid chains associated with the oil or fat which are largely palmitic, stearic, oleic, and linoleic acids for naturally occurring oils and fats (Biodiesel 2009).

2.1.2 The Biodiesel Reaction

Figure 3: Biodiesel Production Process

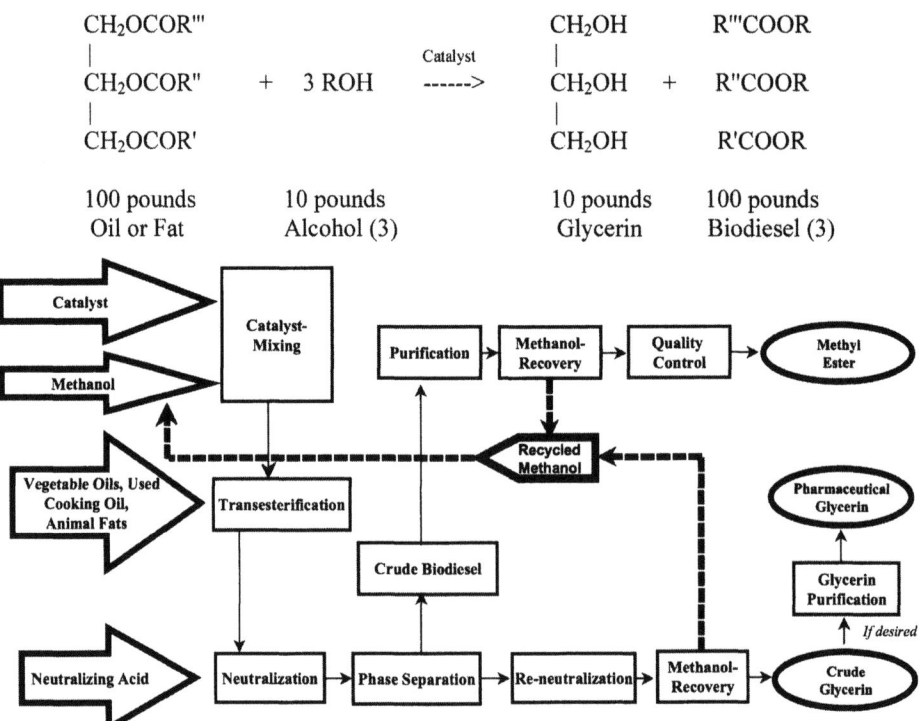

Source: Biodiesel 2009

2.1.3 Bioethanol

All fermentable sugars (i.e. glucose, sucrose, etc.) may be converted to ethanol by fermentation. Such sugars are present in a polymerized state in many species of the plant world such as sugarbeets, sugarcane, wheat, corn, potatoes, but also in grass or wood. Some asset products (e.g. cheese whey, waste paper, etc.) and various residues may also be converted into bioethanol.

Depending on the state of polymerization, the sugars have to undergo one or several treatment steps, with the aim of transforming the various polymer chains in simple fermentable sugars. After fermentation by means of micro-organisms (yeasts, bacteria, etc.), ethanol is recovered by distillation (hydrous ethanol at 92-96% vol.) followed by dehydration (anhydrous ethanol at 99,7% vol.). The figure below shows the schematic process of bioethanol production from cereals (Biofuels-platform 2009).

Figure 4: Production of Bioethanol

1. Unloading
2. Milling
3. Mashing
4. Cooking
5. Hydrolysis
6. Cooling
7. Fermentation
8. Distillation
9. Dehydration
10. Storage
11. Stillage treatment

Source: Biofuels-platform 2009

2.1.4 Biomass

Naturally biomass can include a wide range of materials. The realities of the economics mean that high value materials for which there is an alternative market, e.g. high quality materials and large timber, are very unlikely to become available for energy applications. However there are huge resources of residues, co-products and waste that could potentially become available.

There are five basic categories of material:

- Virgin wood, from forestry, arboricultural activities or from wood processing;
- energy crops: high yield crops grown specifically for energy applications;
- agricultural residues: residues from agriculture harvesting or processing;
- food waste, from food and drink manufacture, preparation and processing, and post-consumer waste;

- industrial waste and co-products from manufacturing and industrial processes (Biomassenergycentre 2009).

2.1.4.1 Conversion technologies

As there is wide diversity in the characteristics and properties of these different classes of material, and their various sub-groups, there is also a wide range of conversion technologies to use these optimally, which include both thermal and chemical conversion technologies (Biomassenergycentre 2009).

There are a number of technological options available to make use of a wide variety of biomass types as a renewable energy source. Conversion technologies may release the energy directly, in form of heat or electricity, or may convert it to another form, such as liquid biofuel or combustible biogas. While for some classes of biomass resources there may be a number of application options, for others there may be only one appropriate technology.

2.1.4.1.1 Thermal Conversion

These are processes in which heat is the dominant mechanism to convert the biomass into another chemical form. The basic alternatives are separated principally by the extent to which the chemical reactions involved are allowed to proceed:

- Combustion;
- gasification;
- pyrolysis.

There is a number of other less common, more experimental or proprietary thermal processes that may offer benefits such as hydrothermal upgrading (HTU) and Hydroprocessing. Some have been developed for use on high moisture content biomass, including aqueous slurries, and allow them to be converted into more convenient forms (Biomassenergycentre 2009).

2.1.5 Syngas

One advantage of the use of syngas to produce fuels is that syngas can be produced from waste materials that would otherwise need to be discarded. Instead of placing waste products in landfills, these waste products can be used to generate a useful, energy rich product. This makes the syngas conversion process both an efficient means of producing energy and an environmentally friendly option for the recycling of waste products.

Another method for producing syngas is the pyrolysis of glycerol. Glycerol was selected as a potential source of syngas because it is estimated that the rising production of biodiesel will result in increasing amounts of glycerol as a byproduct.

Possibly the most promising process for the generation of syngas is the gasification of plant biomass. Virtually any carbon-based material can be gasified to produce syngas. The carbon that is stored in the plant material is released as carbon monoxide and carbon dioxide that makes it possible to convert nearly 100% of the carbon in the biomass to syngas (MMG 445 Basic Biotechnology eJournal D. Mackaluso).

2.2 Common techniques applied in the sector of biofuels production

2.2.1 Biodiesel

Biodiesel is a high energy, organic compound that can be made from various triglycerides (oils), an alcohol and a catalyst through a transesterification reaction. This reaction is preferable because fatty acids are the most common source of reduced carbonchains.

The major issue confronting biodiesel formation today is the presence of water and free fatty acids in most oils. These molecules result in the formation of soap during the transesterification of the triglyceride, which decreases the yield and requires refinement to remove the soap. This is seen especially in large scale operations where an alkali catalyst is utilized. While it is possible to remove these molecules from the oils themselves, it is an expensive process and one which outweighs the benefits of using biodiesel. With current research showing a variety of oils that can be used in this process it becomes apparent that the type of oil used will ultimately dictate the price of biodiesel. Along with the choice of a feedstock, research is also focusing on the selection of a catalyst and the possible use of acceptor molecules. To produce a biodiesel molecule, the oil must undergo transesterification enabling a fatty acid to undergo esterification. Transesterification and esterification of oils and fatty acids can be accomplished in the presence of an alcohol and a catalyst. Common alcohols used in this process are short chain alcohols, most notably methanol and ethanol. The catalysts commonly used for this process are chemical catalysts, biocatalysts, and nonenzymatic heterogeneous catalysts.

One of the major issues of producing biodiesel is limiting the presence of water and free fatty acids in solution. This can result in the formation of soap in the presence of some catalysts. This can be directly linked to the quality of the feedstock used in the transesterification process. The discrepancy in the scientific world is the debate of whether non-refined oils or refined (virgin) oils should be utilized. The higher the grade and the more refined the oil is, the less water it is likely to contain. The drawback here, though, is the high cost of refining such oils. To overcome this hindrance, solvents and other acceptor molecules are being tested to bind to the reactive ends of these molecules, rendering them inactive. Another issue confronting the production of biodiesel is, during the transesterification of oil, glycerol is also formed in a 1:10 ratio. The presence of glycerol may require additional purification and has been shown to reduce the function of some biocatalysts because it can act as a substrate in some reactions.

2.1.1.1 Feedstock

In general, virgin oils tend to have much less water and fewer free fatty acids than those of the unrefined oils but also tend to cost much more to produce. Though non-refined oils are much cheaper to produce, further refinement is required throughout the process. Oils suitable for use as a feedstock should have a free fatty acid content lower than 1% and a water content lower than 0,06%. Currently, the most common feedstocks used are rapeseed oil, sunflower oil, soybean oil and animal fats. Animal fats tend to be low in free fatty acids and water, but there is a limited amount of these oils available, meaning these would never be able to meet the worldwide fuel. There is also ongoing research on the production of suitable lipids produced through microbial growth. This process would utilize the lipids that are produced in molds, yeasts and algae during metabolic stress.

2.2.1.2 Chemical catalysts

Chemical catalysts are the most studied and most often used catalysts during biodiesel formation. This can be accredited to the availability and low cost of these compounds. Chemical catalysts generally proceed at a faster rate than other forms of catalysis. In biodiesel production, there are two classes of chemical compounds used, alkali catalysts and acid catalysts. There is a large amount of research being conducted using these two systems independently, but research has just begun on using these two methods together. One such study has found that increased yield and diminished reaction time occurs when using an acid pretreatment on the oil, followed by the addition of an alkali catalyst.

2.2.1.3 Alkali catalysts

The majority of recent research has been centered on the use of alkali catalysts to mediate the transesterification of fatty acids. This process has shown to be faster than that of the acid catalysts but dictates more moderated reaction conditions. This process requires less alcohol to be used in the reaction than acid catalysts. These reactions however produce soap more commonly than those of acid catalysts when the content of free fatty acids in the oil is high. This occurs because the free fatty acids and water are neutralized more readily by bases than by acids. These reactions then, in turn, consume the catalyst. Because of these soap forming properties, alkali catalysts are inefficient in large commercial operations, where the levels of water and free fatty acids are difficult to regulate.

2.2.1.4 Acid catalysts

Acids used in the catalysis of the transesterification of biodiesels are usually either hydrochloric acid or sulfuric acid. Though these two acids are the most common, any Bronsted acid can be used in this reaction. A large amount of commercially produced biodiesel is synthesized using acid catalysis. For this process, an acid-catalyzed esterification processor is used because of the low price and the ability of acids to minimize the amount of water and free fatty acids involved in the reaction. This method is commercially useful, because the mass production of biodiesel usually results in the presence of a relatively large amount of free fatty acids. The use of an acid catalyst is observed to be more effective than alkali catalysts when the concentration of free fatty acids is high (> 1%).

2.2.1.5 Biocatalysts

Biocatalysts have been becoming increasingly important in the discussion of biodiesel preparation lately. It is even hypothesized that these catalysts will eventually have the ability to outperform chemical catalyst. Biocatalysts are naturally occurring lipases which have been identified as having the ability to perform the transesterification reactions that are essential to biodiesel production. These lipases have been isolated from a number of bacterial species: *Pseudomonas fluorescens*, *Pseudomonas cepacia*, *Rhizomucor miehei*, *Rhizopus oryzae*, *Candida rugosa*, *Thermomyces lanuginosus*, and *Candida antarctica*. According to a recent study, the *Pseudomonas cepacia* bacterium is the most promising organism for producing a lipase that can be used in transesterification (MMG 445 Basic Biotechnology eJournal 2008).

2.2.2 Ethanol

The traditional biological conversion routes for bioethanol production are well established (Figure 5 below). The main raw materials that have to be extracted are sucrose or starch. For sucrose from sugarcane or sugar beet crops, the juices are first mechanically pressed from the cooked biomass followed by fractionation. The sucrose is metabolised by yeast cells fermenting the hexoses and the ethanol is then recovered by distillation. Starch crops must first be hydrolysed into glucose before the yeast cells can convert the carbohydrates into ethanol. Pre-treatment consists of milling the grains of corn, wheat or barley followed by liquefaction and fractionation. Acidic or enzymatic hydrolysis then occurs prior to fermentation of the resulting hexoses. Although highly efficient, the starch grain-based route consumes more energy (and thus potentially emits more CO_2 into the atmosphere depending on the energy sources used), than the sucrose-based route. From the fermentation process onwards, both routes are almost identical. Overall using either sugar or starch is a mature technology to which few significant improvements have been made in recent years (OECD/IEA 2008).

Figure 5: Conversion routes for sugar or starch feedstocks to ethanol and co-products

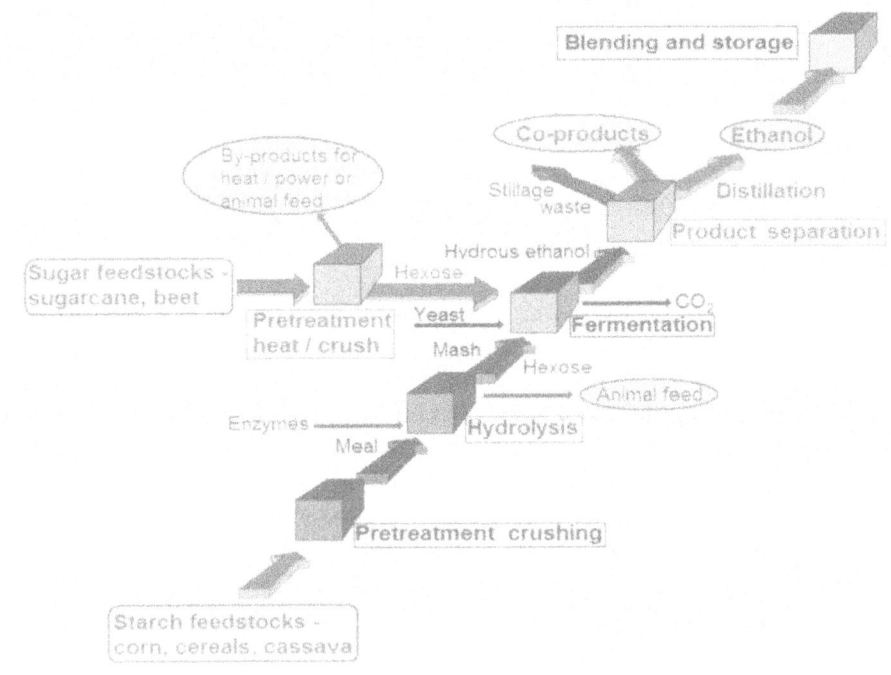

Source: OECD/IEA 2008

Bioethanol is produced through fermentation of sugars by yeasts or bacteria. Mostly used in the ethanol industry is the yeast *Saccharomyces cerevisiae* ('bakers yeast') which is capable of fermenting glucose, fructose, sucrose, galactose and mannose (Roehr 2001). Bioethanol can thus be produced from any feedstock that contains significant amounts of these sugars or glucose polymers such as starch and cellulose that can be converted into glucose via hydrolysis (or 'saccharification'). The main sugars and sugar polymers used for ethanol production are depicted in Figure 6.

Figure 6: Major sugars and sugar polymers for bioethanol production

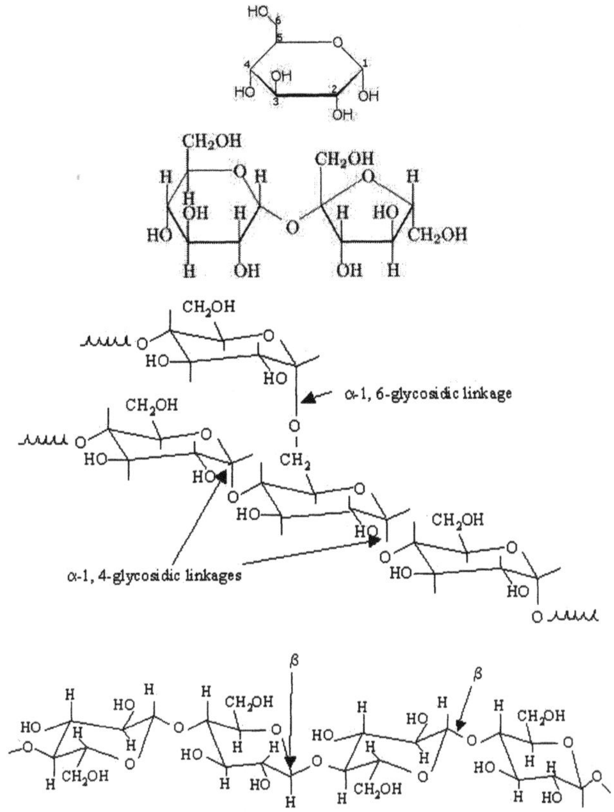

Source: SenterNovem 2009

Sugar obtained from feedstocks such as sugar beets, sugar cane and 'molasses', a by-product from sugar production, can be fermented directly. Starch from agro-feedstocks such as corn, potatoes, wheat, rye, barley and sorghum is a glucose polymer that must be enzymatically hydrolysed to glucose monomers prior to fermentation. Production of fermentable sugars from (ligno) cellulose requires more rigorous pre-treatment due to the strong intermeshment of the sugar polymers (cellulose and hemicellulose) with each other and other biomass components.

2.2.2.1 Pretreatment of sugar crops

For example in tropical regions like e.g. Brazil, sugar and bioethanol is produced from sugar cane, a crop with a high yield, but only suitable for the tropical climate. Sugar beets are more versatile sugar crops, because they can tolerate a wide range of soil and climatic conditions and consequently they can be produced in most European countries. Sugar beets are used for ethanol production only in Europe. The sucrose content of the beets is typically 15-20% of the dry weight. In the processing plant the beets are washed, sliced and passed into a 'diffuser' to extract the sugar into a hot water solution. The liquid exiting the diffuser is called 'raw juice'. At a certain point, further sugar extraction is not economically attractive. The remaining syrup ('molasses') contains 45 wt% sugars, and can be fermented to ethanol. The remaining pulp contains 95% moisture and can be pressed to recover sugar, which is added to the raw juice. The pressed pulp is dried and sold as animal food. Alternatively, sugar syrup may be produced directly from sugar beet by cooking shredded sugar beets for several hours and then pressing the resulting beet mash and concentrating the juice. The raw juice can be used for production of sugar or bioethanol. If the raw juice is used for sugar production, it has to be purified and partly evaporated to produce a concentrated juice, from which the sugar is crystallized (see Figure 7).

Figure 7: Scheme of a combined sugar/bioethanol production process from sugar beet

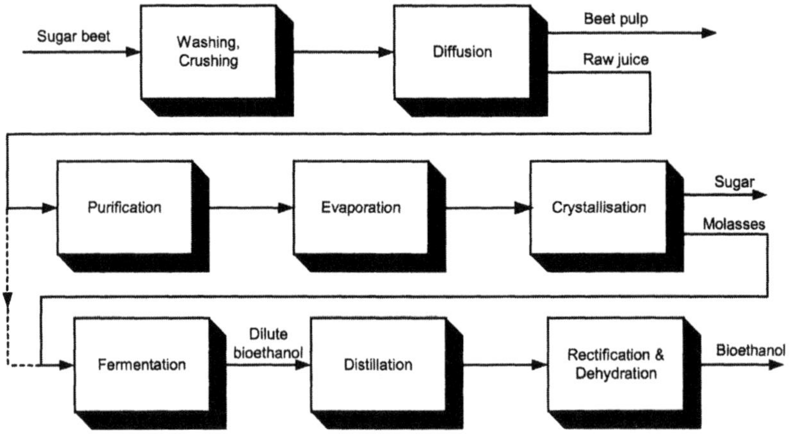

Source: SenterNovem 2009

In Germany and Eastern Europe potatoes are the most widely used starch source. Wheat, rye and barley grains typically contain 60-70 wt% starch, 15 wt% water; while the remainder consists of mainly proteins, but also some fats, cellulose and minerals. Fresh potatoes generally contain 75 wt% water and approximately 17 wt% starches, with the remainder being proteins, small amounts of sugars and other compounds. Ethanol processes based on starch are more complicated than those using sugars directly, because the starch has to be hydrolyzed to glucose prior to fermentation. The most common process used in Europe is the 'milling and mashing process at higher temperatures'. In this process, first starch is released from the cell material ('liquefaction') and then the starch is converted to fermentable sugars ('saccharification') by addition of enzymes (amylases). The process can use all starch containing raw materials. After washing the raw materials to remove sand, stones, soil etc., they are reduced in size, generally to 1,5 mm with a hammer mill. Preheated water and liquefaction enzymes are added and the mash is heated with steam to 65-95°C, depending on the type of feedstock. The liquefaction takes 30 minutes to several hours. The mash is cooled down for saccharification and cooled down further to the temperature required for fermentation. The performance of this process depends on the efficiency to break up cells during milling and on the efficiency of the enzymes used. The process can be executed as a batch or as a continuous process.

Figure 8: Scheme for bioethanol production from starchy raw materials

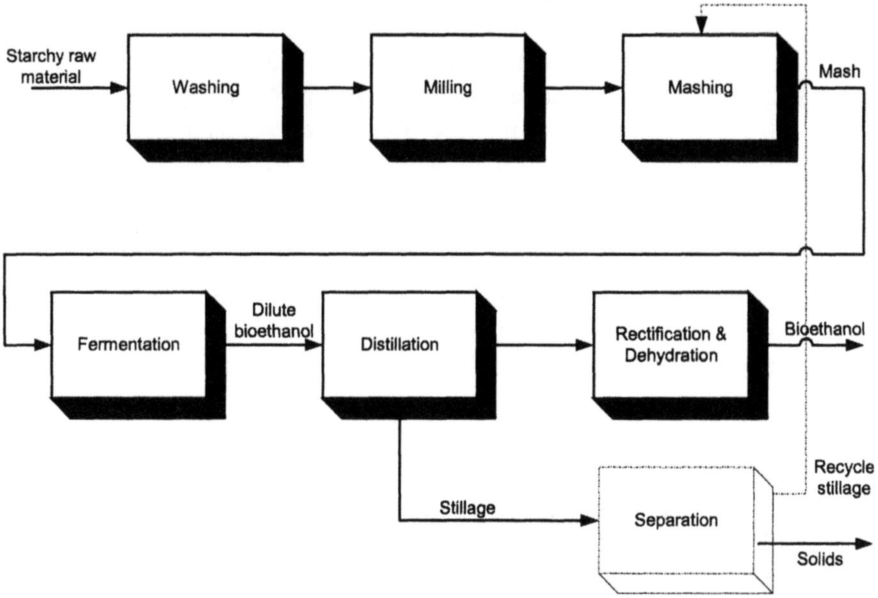

Source: SenterNovem 2009

After the starch is converted to glucose, the mashes are fermented to ethanol. The fermentation product 'beer', containing ethanol and solids, is distilled. The water/solids mixture obtained after distillation ('stillage') can be used as animal feed or fertilizer, either in liquid or dried form. The water can be recycled after the solids are mechanically removed, in order to reduce water and energy demand of the production process.

Industrial fermentation of sugar to ethanol is generally performed with the yeast *Saccharomyces cerevisiae* at temperatures of 28-35°C. The fermentation process can only be executed in relatively dilute concentrations, because yeasts are susceptible to ethanol inhibition. Concentrations in the order of 10-20% may completely halt yeast growth in batch fermentations. Continuous processes where yeast is added allow higher ethanol concentrations up to 20%. Certain species of bacteria can also ferment sugar to ethanol, but they often also produce by-products, such as organic acids. Furthermore, infections are more difficult to control in bacterial fermentations. The fermentation takes place in large cylindrical fermentors, generally in a batch process, for periods of 10-60 hours. Following fermentation the yeast and other solids are often separated from the 'beer' by centrifugation, and may be recycled to the fermentor. For further information to fermentation processes please consider the working paper on Biogas produced within the RENET project.

2.2.2.2 Distillation and final dehydration

Dedicated engines that can use 95% ethanol ('hydrous ethanol') are currently used only in South America. In order to be used as a component in blends with petrol, bioethanol has to be purified to more than 99,5% purity. A first distillation or stripping column removes ethanol from the "beer" giving an approximately 50/50 water/ethanol mixture. Remaining solids are removed in the 'bottoms' or 'stillage' fraction, if these were not already removed prior to distillation. A second column (rectifier) removes water up to 95% ethanol. Higher alcohols or "fusel oils" are also removed in this column. The limit for distillation is 95-96% ethanol due to the water/ethanol azeotropic system. Therefore, the remaining water has to be removed with a different technique, such as dehydration with molecular sieves. Distillation requires a large amount of energy. Increasing energy costs, especially in Europe, have led to an increased emphasis on heat recovery and reduction of energy use. In practice this is realised by using vapor recompression systems or multiple-stage, high-pressure distillation systems, resulting in a 40-80% reduction of steam consumption. The investment costs for these systems are higher than for conventional distillation technology. The actual energy usage of the distillation depends strongly on the ethanol concentration after the fermentation process. The higher the ethanol concentration, the lower the energy cost for the distillation. Final dehydration to 99,5% bioethanol requires additional energy (SenterNovem 2009).

2.2.2.3 Second Generation Biofuels

Projections for second generation fuels to become commercial are wide ranging but often considered unlikely to occur before 2015 (IEA 2008a). The basic conversion technologies are not new and their commercial development has been a long time coming – successful development is not yet guaranteed. Considerable investment in pilot and demonstration plants has been made worldwide but how and when commercial scale-up can be realised is the key question.

Second generation biofuels are expected to be superior to many of the first generation in terms of the energy balances, greenhouse gas emission reductions, land use requirements, and competition for land, food, fiber and water. However they do not produce co-products such as animal feeds which should also be considered in a comparison (Renewable Fuels Agency 2008).The main reason why they have not yet been taken up commercially, despite their potential advantages over first generation biofuels, is that the necessary conversion technologies (from feedstock to finished fuel) are not technically proven at a commercial scale and their costs of production are estimated to be significantly higher than for many first generation biofuels at the moment. Further research is required on land use requirements, effects of co-products, water use and energy for processing as outlined below.

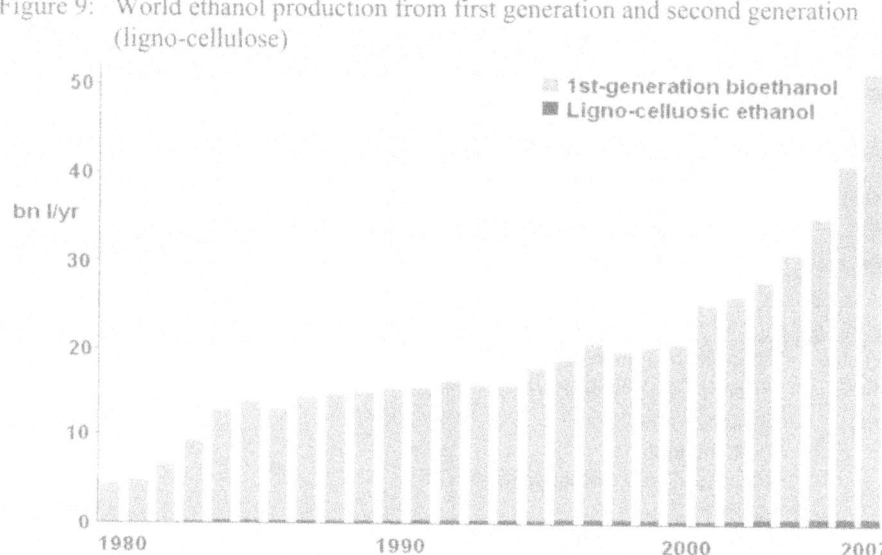

Figure 9: World ethanol production from first generation and second generation (ligno-cellulose)

Source: Mabee and Saddler 2007

Second-generation biofuels can be broadly grouped into those produced either biochemically or thermo-chemically, either route using non-food crops, especially from ligno-cellulosic feedstocks sourced from crop, forest or wood process residues, or purpose-grown perennial grasses or trees. Such crops are likely to be more productive than most crops used for first generation in terms of the energy content of biofuel produced annually per hectare (GJ/ha/yr).

2.2.2.3.1 Ligno-cellulosic feedstocks

Concerning its composition, ligno-cellulose is the botanical term used for biomass from woody or fibrous plant materials, being a combination of lignin, cellulose and hemicellulose polymers interlinked in a heterogeneous matrix. The relative importance of each of the polymers can vary significantly with the feedstock type. The combined mass of cellulose and hemicellulose in the plant material varies with species but is typically 50-75% of the total dry mass with the remainder consisting of lignin. Cellulose is a straight chain polymer consisting of units of glucose (a 6 carbon (C6) sugar) less one molecule of water connected via specific linkages so that each link has a formula $C_6H_{10}O_5$. Hemicellulose is a heterogeneous material which in agricultural and woody substrates is primarily a polymer of predominantly Xylose and Arabinose (both pentoses, being C5 sugars), combined with three different Hexoses (C6). Lignin is composed of a number of phenolic compounds that may act as an inhibitor to the hydrolysis or fermentation of sugars so its presence creates challenges for bioconversion processes.

In the biochemical conversion process that relates to the concept of bio-refineries, lignin represents a potential valuable source of chemical feedstock. In ethanol plants it may be combusted to provide process heat and power. In the thermo-chemical route, all polymers, including lignin, are converted to synthesis gas. Material acceptance and registration.

Table 1: Classification of Second Generation Biofuels

Biofuel group	Specific biofuels	Biomass feedstock	Production process
Bioethanol	Cellulosic ethanol	Ligno-cellulosic materials	**Advanced enzymatic hydrolysis and fermentation
Synthetic Biofuels	Biomass to liquids (BTL), Fischer-Tropsch diesel (FT) Synthetic diesel Biomethanol Heavier alcohols (butanol and mixed) Dimethyl ether (DME) P-Series* (ethanol + MTHF) etc	Ligno-cellulosic materials	***Gasification and synthesis
Biodiesel (hybrid of first and second)	NExBTL H- Bio Green pyrolysis diesel* Algal oil*	Vegetable oils and animal fats Ligno-cellulosic materials Algae	Hydrogenation (refining) ***Pyrolysis Cultivation
Methane	Bio synthetic natural gas* (SNG)	Ligno-cellulosic materials	***Gasification and synthesis
Bio hydrogen	Hydrogen	Ligno-cellulosic materials	***Gasification and synthesis of **biological processes

* *Some fuels listed can be classified as "advanced" biofuels*
** *Bio-chemical route: After comminution of the biomass feedstock and pre-treatment, ethanol can be produced by the hydrolysis of lingo-cellulosic raw materials, the fermentation of the extracted sugars followed by distillation and formulation to give the final fuel product. Fermentation of the glucose sugars is mature commercial technology, but the hydrolysis of agricultural residues and woody biomass and the fermentation of pentose still need further development*
*** *Thermo-chemical route: – indirect liquefaction methods require the biomass to be the first pyrolysed to boil-oil, or gasified and the product gas cleaned and processed to form synthesis gas (mainly CO and H2). This gaseous mixture can then be used in a commercial chemical process to synthesise a range of liquid biofuels including methanol, Fischer – Tropsch diesel, DWE (dimethyl ether) or as gaseous methane or hydrogen fuels*

Source: OECD/IEA 2008

2.2.2.3.2 Technological conversion routes

Several technological conversion routes exist for producing second generation liquid or gaseous biofuels from solid biomass. However, none have yet reached the fully commercial stage; hence no clear technology leader or pathway has emerged. The bio-refinery concept, usually based on either thermo- or bio-chemical routes, is where biofuels are produced from single or multifeedstocks along with one or more co-products, as well as possibly heat and power produced for use on site and/or for export. The concept of producing small quantities of high value products (e.g. chemicals) and larger quantities of low value products (e.g. biofuels) theoretically maximizes returns from the biomass feedstock by improving economic performance in the same way that oil refineries do for crude oil today (OECD/ IEA 2008).

Figure 10: Ethanol production from lingo-cellulose vie the bio-chemical route

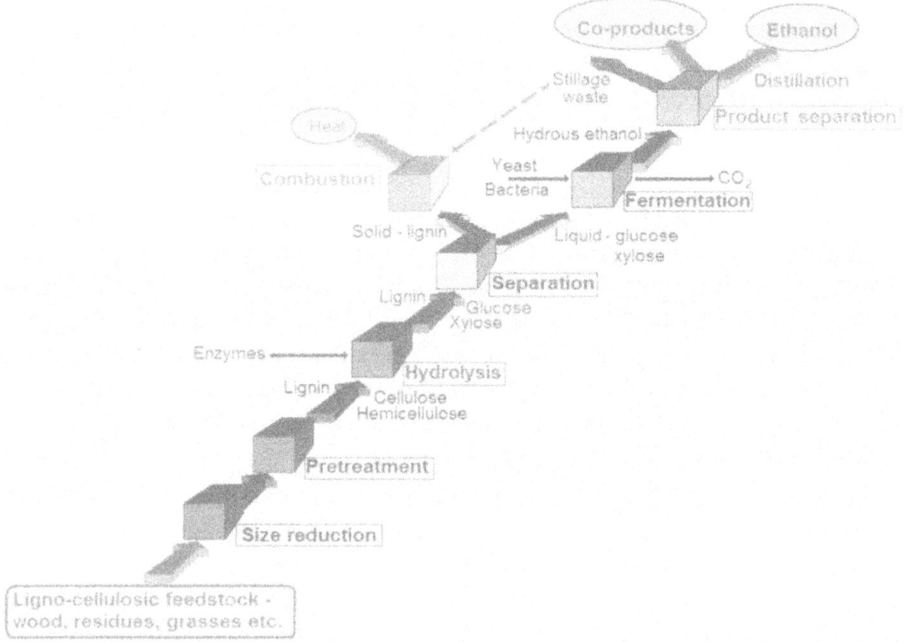

Source: OECD/IEA 2008

Bio-chemical conversion uses biological agents, specifically enzymes or microorganisms, to carry out a structured deconstruction of the ligno-cellulose into its base polymers and to further break down cellulose and hemicellulose into monomeric sugars including glucose and xylose. These sugars can then be fermented into ethanol. Feedstock is based upon agricultural and forest biomass (either residues or dedicated crops) but could also include the potential recovery of biomass from urban municipal solid waste (MSW) streams (OECD/IEA 2008).

2.3 Solid Biofuels

Solid biofuels include the following sources:

- Products from agriculture and forestry;
- vegetable waste from agriculture and forestry;
- vegetable waste from food processing industry;
- wood waste, with the exception of wood waste which contains halogenated organic compounds or heavy metals as a result of treatment with wood preservatives or coating, and which includes in particular such wood waste from construction- and demolition waste;
- cork waste;
- fibrous vegetable waste from virgin pulp production and from production of paper from pulp, if it is co-incinerated at the place of production and heat generated is recovered.

2.3.1 Wood Pellets

Pellets are the solid biofuel that has the highest refining degree. In the same way as briquettes they are compressed chips, but based on a more finely ground raw material and with lengths smaller than 25 mm. Standard diameters are 6, 8 and 12 mm. Pellets are suitable for smaller plants and are normally used up to 1 MW, but in some cases pellets are also used in larger plants. Pellets have properties similar to those of oil with regard to transport, storage and combustion control. Oil-fired plants can often be converted relatively easily for pellet firing (Renewableen 2007).

Wood pellets are made by compressing sawdust and wood shreds produced in sawmilling or manufacturing. Pellets are compact and well suited to automated feed systems because they are flowable. They also have a high energy value because of their density and low moisture content (typically 8-10%). The energy density of pellets by volume is also more consistent than logs or chips, making them an easily tradable commodity. From an energy balance perspective pellet production is usually only justifiable if the sawdust raw material is a by-product of another process (Forest Research 2006).

Figure 11: Wood pellets production process

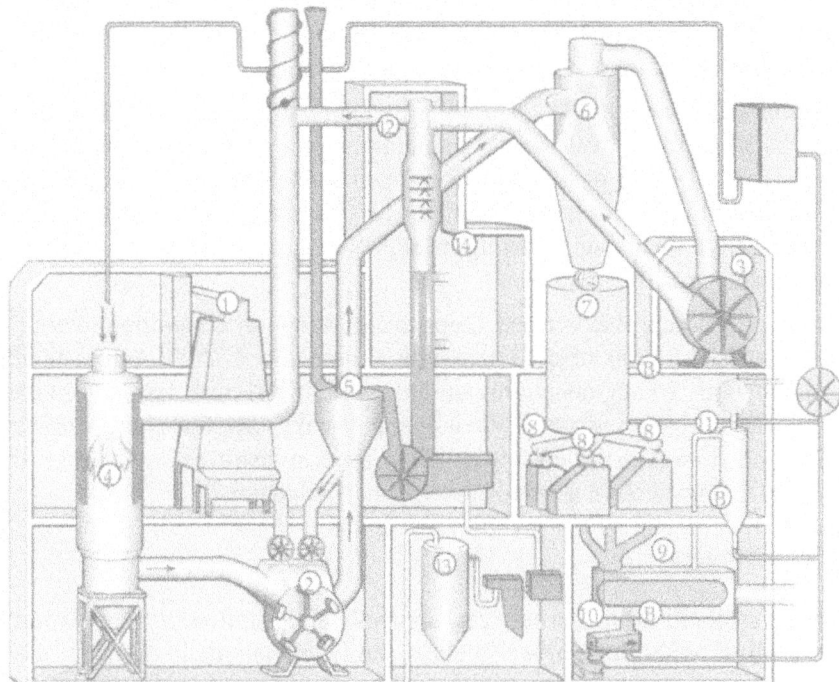

Source: SCA BIONORR AB 2009

The unprocessed wood products are conveyed from a loading and buffer bin (1), into the mill (2) where they are ground into a fine powder. Flue gases from a hot-gas generator (4) are used to dry the material. After the milling and drying is complete, the powder is blown through the drying system by a fan (3) to a screen (5) where the oversized powder is collected and recirculated to the mill. The dried powder is separated from the hot gases and water vapor in the cyclone separator (6). It is then stored in a buffer silo (7) that feeds the pellet machines (8). The pellet machines press the powder through a die to form pellets that are now at approximately 100°C. The hot pellets are cooled in the air cooler (9) until they have reached a suitable storage temperature. Prior to storage the remaining finer particles are removed in a vibration separator (10). The dust that is removed with the air cooling (11) and the fine particles removed by the vibration separator (10) are recycled as fuel (B) for the hot-gas generator. The main part of the hot gases and steam produced during the drying process are returned to the hot-gas generator for reheating. Small amounts of gas and steam are released after passing through a smoke scrubber (12, 13). During the scrubbing process, the steam is condensed in a heat exchanger (14). The resulting heat is e.g. transferred to the municipal district-heating network (SCA Bionorr AB 2009).

2.3.2 Wood Chips, Demolition Wood, Briquettes

Wood chips and chips from demolition wood can have a varying degree of refinement. Chip quality will therefore depend on the type of wood, the equipment used for making the chips, sorting techniques and moisture content. Wood chips can be used in all plant sizes, but chips as fuel will normally require more attention and total investment compared to more refined biofuels. Dry chips are a fuel that can be stored, but moist chips start to compost if left too long.

Demolition wood and treated wood will often end up as a waste fraction that can only be burned in approved waste incinerators. If wood waste is not contaminated, for example with paint, sealants or chemicals used as preservatives, it can be refined to chips or briquettes that can be used in normal incinerators. This fuel is normally crushed and passed through a sieve, as opposed to wood chips, which are chopped.

Briquettes are compressed, dried chips from wood or demolition wood. The chips are pressed to logs or cylinders with a diameter of 25-70 mm. The length varies up towards 20 cm, depending on the raw material qualities and production processes. Briquetting reduces the volume and makes the fuel more suitable for transport and storage. Briquettes are mainly used in heating plants larger than 1 MW, but also burn well in a wood stove (RenewableenergyNor 2009).

2.4 Small Boilers

Distinctions should be made between manually fired boilers for fuel wood and automatically fired boilers for wood chips and wood pellets. In order for the manual boilers not to need feeding at intervals of 2-4 hours a day, during the coldest periods of the year, the boiler's nominal output is selected so as to be up to 2-3 times the output demand of the dwelling. Automatic boilers are equipped with a silo containing wood pellets or wood chips. A crew feeder feeds the fuel simultaneously with the output demand of the dwelling.

Great advances have been made over the recent 10 years for both boiler types in respect of higher efficiency and reduced emission from the chimney (dust and CO). Improvements have been achieved particularly in respect of the design of combustion chamber, combustion air supply, and the automatics controlling the process of combustion. In the field of manually fired boilers, an increase in the efficiency has been achieved from below 50% to 75-90%. For the automatically fired boilers, an increase in the efficiency from 60% to 85-92% has been achieved (Videncenter 2004).

2.4.1 Manually Fired Boilers

The principal rule is that manually fired boilers for fuelwood only have an acceptable combustion at the boiler rated output (at full load). At individual plants with oxygen control, the load can, however, be reduced to approx. 50% of the nominal output without thereby influencing either the efficiency or emissions to any appreciable extent. By oxygen control, a lambda probe measures the oxygen content in the flue gas, and the automatic boiler control varies the combustion air inlet. In order for the boiler not to need feeding at intervals of 2-4 hours a day, during the coldest periods of the year, the fuel wood boiler nominal output is selected so as to be up to 2-3 times the output demand of the dwelling designed for fuel wood should always be equipped with storage tank. This ensures both the greatest comfort for the user and the least financial and environmental strain. In case of no storage tank, an increased corrosion of the boiler is often seen due to variations in water and flue gas temperatures, and in addition to that, the manufacturer warranty may also lapse.

Figure 12: Determination of storage tank size

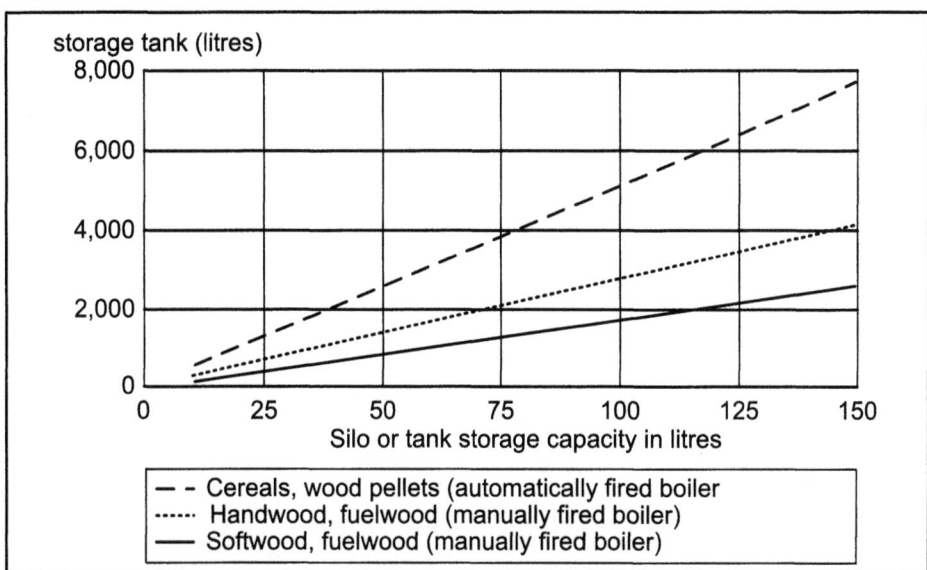

When knowing the boilermagazine size (i.e. the unit of the boiler that is filled with fuelwood), the necessary size of the storage tank can be determined.

Source: Videncenter 2004

2.4.2 Automatically Fired Boilers

Despite an often simple construction, most of the automatically fired boilers can achieve an efficiency of 80-90% and a CO emissions of approx. 100 ppm (100 ppm = 0,01 volume%). For some boilers, emissions reach up to 92% and 20 ppm, respectively. An important condition for achieving these good results is that the boiler efficiency during day-to-day operation is close to full load. For automatic boilers, it is of great importance that the boiler nominal output (at full load) does not exceed the max. Output demand in winter periods (Videncenter 2004).

2.4.3 District Heating Plants

The term district heating plants refers to plants with own generation of heat, but without power generation. The heat is distributed to a district heating system to which all consumers living within the system have the opportunity of being connected.

Figure 13: Thyborøn district heating system

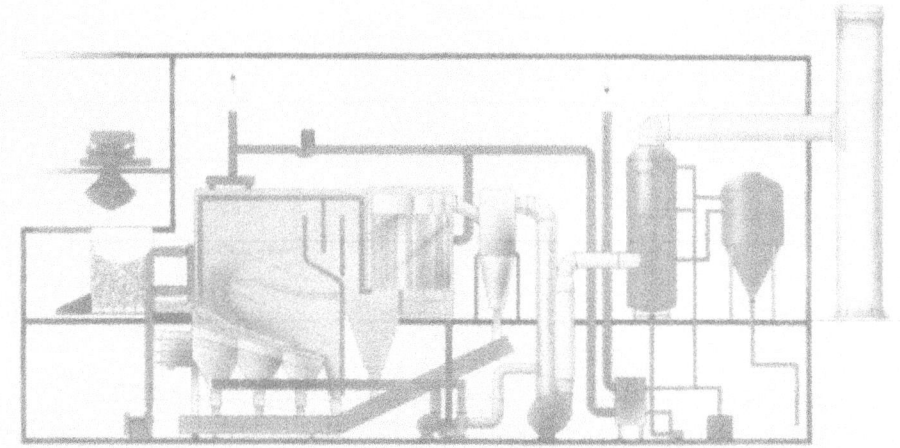

In Thyborøn the district heating is supplied by a 4 MW chip-fired boiler. The system flue gas condenser produces an additional 0,8 MW heat at 50% moisture contained in the wood chips.

Source: Videncenter 2004

Seen in an international perspective, the use of wood chips at district heating plants has increased tremendously during a relative short period of time. Wood chip-fired district heating plants are established either in order to replace oil- or coal-fired district heating plants, connected to old district heating systems, or as new plants and systems (the so-called "urbanization" projects). Wood chip-fired boilers at for example Danish district heating plants are designed for the generation of heat in the range of 1 MW and 10 MW; the average being 3,5 MW (Videncenter 2004).

3. Emissions and Consumption

Depending on the type of biofuel, the production of biofuels has different impacts on the environment. Fuels like biodiesel, bioethanol and syngas require very high energy input per kg output. Their sophisticated processing also means that additional actions are required in order to keep their impact on the environment as small as possible. The solid biofuels require less preprocessing, they are easier for transportation what in consequence has a smaller environmental impact compared to the liquid biofuels. Emissions strongly depend on the type of biofuel and the according production process and cannot be evaluated in this working paper.

4. Techniques to consider in the determination of Best Available Techniques

Additional techniques are considered generally to have potential for achieving a high level of environmental protection in the biofuels production sector. For example management systems, process-integrated techniques for process measurement have significant influence on plants efficiency and therefore on the environment. Prevention, plant control, design, management and product utilization procedures as well as material collection systems and efficient energy usage must be considered to achieve the objectives of IPPC. Because it is not possible to be exhaustive and because of the dynamic nature of biofuels production industry, it is possible that there may be additional techniques left unmentioned, but which may also be considered as Best Available Techniques (BAT).

5. Best Available Techniques

Best Available Techniques in the biofuels production sector are highly dependant on the targeted product, the product quality, the availability of raw materials, the geographical location and its specific features.

Depending on the kind of biofuel production BAT is the use of aforementioned techniques (see 2 Applied Techniques) in such way that the biofuel (biodiesel, bioethanol, syngas etc) production rate reaches the maximum and that emissions to air, water and soil are minimal. It is not possible to define BAT for a single biofuel production plant, because all the techniques depend on bases of operations as there are kind of input materials that are used, the location of the plant, size of the plant, energy demand, etc.

6. Emerging Techniques

The transition from a fossil fuel-based economy to a biofuel-based economy depends on the improvement of biofuel yields via the successful development of suitable microorganisms capable of efficiently fermenting a variety of sugars while simultaneously displaying tolerance to high end-product concentrations. Cellulose (third generation Biofuel) represents an attractive feedstock for biofuels production because of its abundance, cost and ability to efficiently be degraded by cellulolytic bacteria.

There will not be "the biorefinery", because different approaches have their relevance depending for example on availability of certain feedstock and process technology and changes in market situations. Today two tendencies in the development of biorefineries can be seen:

- Improvement of existing plants with the goal of extending the feedstock and product line-up;
- conception of new, fully integrated biorefineries for a great variety of biomasses and different products from scratch.

In order to develop integrated plants that are technically available, economically feasible and ecologically beneficial, both development directions may increasingly be joined to offer solutions (DBFZ 2008).

The confluence of developments in plant biology and biotechnology, in carbon capture and storage techniques and in innovative bioconversion methods makes it possible to begin to imagine a 'fourth generation' of biofuels and bioenergy systems. The first steps towards such fuels are already being taken. Major research organisations have found that over the long term, there is vast potential for sustainably produced bioenergy. Scientists working for the IEA's Bioenergy Task 40 put it at a maximum of around 1.300 Exajoules (Ej) by 2050 (current global fossil fuel use is around 380Ej per year). This biomass potential is explicitly based on a 'no deforestation' scenario and on the fact that all food, fiber and fodder needs of growing populations and livestock must be met first. After taking these requirements into account, the researchers find vast potential especially in Africa (320Ej) and Latin America (220Ej).

High potentials can be seen in the breakthroughs in biotechnology, such as the design of high yielding dedicated energy crops. Developments in this field are going on very rapidly. High biomass crops, trees with increased carbon storage capacity, drought tolerant energy crops, grass species that beat the major problem of acidic soils, new plants with particular properties catering to a specific bioconversion process (e.g. low lignin trees, maize with embedded enzymes for rapid conversion) have already shown what could be possible in future. The combination of such crops with advanced bioconversion techniques that allow the capture and storage of carbon dioxide makes it possible to yield a 'fourth generation' of biofuels (Biopact 2009).

7. References

(AmberWaves Nov. 2007)	*Amber Waves* is published four times per year (March, June, September, and December) by the U.S. Department of Agriculture, Economic Research Service. http://www.ers.usda.gov/AmberWaves/November07/Features/Biofuels.htm, last visit: 10 June 2009
(EurObserv 2008)	EurObserv 2008 – The State of Renewable Energies in Europe 8th EurObserv'ER Report, last visit: 10 June 2009
(The Wharton School of the University of Pennsylvania 2007)	"Economic and Business Challenges for Biodiesel Production in Turkey" Paul R. Kleindorfer, http://opim.wharton.upenn.edu/risk/library/2007_PRK-UGO_BiodieselTurkey.pdf, last visit: 10 June 2009
(Biodiesel 2009)	National Biodiesel Board, http://www.biodiesel.org/pdf_files/fuelfactsheets/prod_quality.pdf, last visit: 10 June 2009
(Biofuels-platform 2009)	http://www.biofuels-platform.ch, http://www.biofuels-platform.ch/en/infos/bioethanol.php, last visit: 10 June 2009
(Biomassenergycentre 2009)	The BIOMASS Energy Centre (BEC) is owned and managed by the UK Forestry Commission, via Forest Research, its research agency. http://www.biomassenergycentre.org.uk/portal/page?_pageid=76,15049&_dad=portal&_schema=PORTAL, last visit: 10 June 2009
(MMG 445Basic Biotechnology eJournal 2007)	"The use of syngas derived from biomass and waste products to produce ethanol and hydrogen" by Joshua D. Mackaluso, MMG 445 Basic Biotechnology eJournal 2007 3: 98-103 www.msu.edu/course/mmg/445/, last visit: 10 June 2009
(MMG 445 Basic Biotechnology eJournal 208)	A review of the processes of biodiesel production – Michael Sheedlo MMG 445 Basic Biotechnology e-Journal 2008 4:61 – 65 http://ejournal.vudat.msu.edu, last visit: 10 June 2009
(OECD/IEA 2008)	Agency IEA Bioenergy and Jack Saddler, Warren Mabee© OECD/IEA, November 2008 From First to Second Generation Biofuel Technologies: An overview of current industry and RD&D activities Ralph Sims, Michael Taylor International Energy

(SenterNovem 2009)	SenterNovem is an agency under the Ministry of Economic Affairs. Bioethanol in Europe Overview and comparison of production processes Rapport 2GAVE0601 http://www.senternovem.nl/mmfiles/ECNGAVEbioethanoleindrapport_tcm24-280156.pdf, last visit: 10 June 2009
(SCA BIONORR AB 2009)	http://bionorr.se/dokument/Bionorr_eng.pdf, last visit: 10 June 2009
(Renewable-energyNor 2009)	http://www.renewableenergy.no/sitepageview.aspx?articleID =177#anker_ last visit: 10 June 2009
(EU 2003)	Directive 2003/30/EC of the European Parliament and of the Council of 8 May 2003, http://ec.europa.eu/energy/res/legislation/doc/biofuels/en_final.pdf
(Videncenter 2004)	http://www.videncenter.dk/Groenne%20trae%20haefte/Groen_Engelsk/Kap_07.pdf
(IEA Bioenergy 2008)	2008 IEA Bioenergy: Task 39 'Commercializing First and Second Generation Liquid Biofuels from Biomass http://www.task39.org/About/Definitions/tabid/1761/language/en-US/Default.aspx, last visit: 10 June 2009
DBFZ 2008	Workshop: Biofuels and bio-based chemicals Trieste 18-20 September 2008. – "Second and Third Generation of Biofuels and Biorefineries – Considerations and concepts" www.ics.trieste.it/VideoStore/BioWorkshop/2008.09.18_16.00-16.59/presentations/7_Thraen.ppt last visit: 10 June 2009
Biopact 2009	Webpage, Biopact " Organisation of European and African citizens", http://news.mongabay.com/bioenergy/2007/10/quick-look-at-fourth-generation.html, last visit: 17 June 2009

7.1 Figures

Figure 1	Global biofuel production between 2000 and 2007 (IEA/FO Licht)
Figure 2	Worldwide Biofuels Production (FO Licht)
Figure 3	Biodiesel Production Process (Biodiesel 2009)
Figure 4	Production of Bioethanol (Biofuels-platform)

Figure 5	Conversion routes for sugar or starch feedstocks to ethanol and co-products (OECD/IEA 2008)
Figure 6	Major sugars and sugar polymers for bioethanol production (SenterNovem)
Figure 7	Scheme of a combined sugar/bioethanol production process from sugar beet (SenterNovem)
Figure 8	Scheme for bioethanol production from starchy raw materials (SenterNovem)
Figure 9	World ethanol production from first generation and second generation (ligno-cellulose) (Mabee & Saddler 2007)
Figure 10	Ethanol production from lingo-cellulose vie the bio-chemical route (OECD/IEA 2008)
Figure 11	Wood pellets production process (SCA BIONORR AB)
Figure 12	Determination of storage tank size (Videncenter 2004)
Figure 13	Thyborøn district heating system (Videncenter 2004)

7.2 Tables

Table 1	Classification of second generation Biofuels (OECD/IEA 2008)

8. Glossary

8.1 Terms and definitions

Biofuels means liquid or gaseous fuel for transport produced from biomass (EU 2003).

Biomass means the biodegradable fraction of products, wastes and residues from agriculture (including vegetal and animal substances), forestry and related industries, as well as the biodegradable fraction of industrial and municipal waste (EU 2003).

Chapter 5

Best Available Techniques – Working Paper – Photovoltaic

Kerstin Kuchta, Marko Gehrmann, Konstantin Haker

The structure of this document is derived from an original reference document on Best Available Techniques (BAT). The determination of BAT is used as a tool to display the state-of-the-art of photovoltaic technologies with the generally accepted guidelines of the IPCC Directive 2008(1/EC).

Interpretation of Best Available Techniques

The European IPPC Bureau exists to catalyze an exchange of technical information on Best Available Techniques under the IPPC Directive 2008/1/EC and to create Best Available Techniques REFerence Documents (BREFs), which must be taken into account when the competent authorities of Member States determine conditions for IPPC permits. IPPC will apply to a wide range of industrial activities and the objective of the information exchange exercise is to assist the efficient implementation of the directive across the European Union. The BREFs will inform the relevant decision makers about what may be technically and economically available to industry in order to improve their environmental performance and consequently improve the whole environment.[1]

The following definitions have been applied:

The term "Best Available Techniques" as defined in article 2(11) of the IPPC Directive as

> "the most effective and advanced stage in the development of activities and their methods of operation which indicate the practical suitability of particular techniques for providing in principle the basis for emission limit designed to prevent and, where that is not practicable, generally to reduce emissions and the impact on the environment as a whole."[2]

---

[1] Http://eippcb.jrc.ec.europa.eu.
[2] Definition of Best Available Techniques by the European Union in Directive 96/61/EC of 24 September 1996 concerning integrated pollution prevention and control.

On this definition following explanation of BAT is accepted[3]:

B *'Best'* in relation to techniques, means the most effective in achieving a high general level of protection of the environment as a whole.

A *'Available* techniques' means those techniques developed on a scale which allows implementation in the relevant class of activity under economically and technically viable conditions, taking into consideration the costs and advantages, whether or not the techniques are used or produced within the State, as long as they are reasonably accessible to the person carrying on the activity.

T *'Techniques'* includes both the technology used and the way in which the installation is designed, built, managed, maintained, operated and decommissioned.[4]

At the installation/facility level, the most appropriate techniques will additionally depend on local factors. A local assessment of the costs and benefits of the available options may be needed to establish the best regional option. The best choice for energy generation technologies may be justified on:

- local public power supply and distribution network;
- geographical location;
- local environmental considerations;
- meteorological conditions.

The overall objective to implement a maximum level of protection for the environment as a whole will often involve the consideration of different environmental impacts on a local basis. However, the obligation to ensure a high level of environmental protection using renewable energies to reduce carbon dioxide emission to the atmosphere implies that the most appropriate techniques cannot be set on the basis of solely local considerations.

Furthermore the IPPC contains a list of considerations to be taking into account generally or in specific cases when determining BAT keeping in mind the likely costs and benefits of a measure and the principles of precaution and prevention[5]:

---

[3] Http://www.Europa.eu/scadplus/leg/en/lvb/l28045.htm.

[4] Reference Document on Best Available Techniques for Management of Tailings and Waste – Rock in Mining Activities (July 2004) By the Directorate – General JRC Institute for Prospective Technological Studies, sustainability in Industry, Energy and Transport, European IPPC Bureau of the European Commission.

[5] Reference Document on Best Available Techniques for Management of Tailings and Waste – Rock in Mining Activities (July 2004) By the Directorate – General JRC Insti-

(i)   the use of low-waste technology,
(ii) the use of less hazardous substances,
(iii) the furthering of recovery and recycling of substances generated and used in the process and of waste, where appropriate,
(iv) comparable processes, facilities or methods of operation, which have been tried with success on an industrial scale,
(v)   technological advances and changes in scientific knowledge and understanding,
(vi) the nature, effects and volume of the emissions concerned,
(vii) the commissioning dates for new or existing activities,
(viii) the length of time needed to introduce the Best Available Techniques,
(ix) the consumption and nature of raw materials (including water) used in the process and their energy efficiency,
(x)   the need to prevent or reduce to a minimum the overall impact of the emissions on the environment and the risks to it,
(xi) the need to prevent accidents and to minimize the consequences for the environment, and
(xii) the information published by the Commission of the European Communities pursuant to any exchange of information between Member States and the industries concerned on Best Available Techniques, associated monitoring, and developments in them, or by international organisations, and such other matters as may be prescribed.

As a part of the RENET project the scope of BAT for Photovoltaic's covers economical, ecological, scientific and statutory considerations for the EU and Turkey.

1. General Information

1.1 Renewable Energies in Europe

Due to the increase of global primary energy consumption, the world population and the reached climate change, industrial countries started to become accustomed to the changing energy situation and opportunities. The European development of the national participation of renewable energies since 2001 is summarized in table 1.

---

tute for Prospective Technological Studies, sustainability in Industry, Energy and Transport, European IPPC Bureau of the European Commission.

Table 1: Percentage of renewable Energy from total Energy consumption

	2001	2003	2005	Target 2010[a]
Belgium	1,3%	1,6%	2,2%	13%
Denmark	12,3%	14,9%	17,0%	30%
Germany	3,9%	4,4%	5,8%	18%
Finland	27,9%	26,7%	28,5%	38%
France	10,9%	9,9%	9,5%	23%
Greece	6,5%	7,2%	7,5%	18%
Ireland	2,2%	2,2%	3,0%	16%
Italy	5,2%	4,4%	4,8%	17%
Luxembourg	0,7%	0,8%	0,9%	11%
Netherlands	1,6%	1,8%	2,4%	14%
Austria	25,8%	21,8%	23,0%	34%
Portugal	20,5%	21,5%	17,0%	31%
Sweden	40,0%	33,9	40,8%	49%
Spain	9,1%	9,4%	7,6%	20%
UK	0,9%	1,0%	1,3%	15%
EU-15	**7,8%**	**7,5%**	**8,0%**	
Estonia	15,3%	14,9%	18,0%	25%
Latvia	34,4%	31,9%	35,5%	42%
Lithuania	15,3%	15,4%	15,0%	23%
Malta	0,0%	0,0%	0,0%	10%
Poland	6,9%	7,0%	7,2%	15%
Slovakia	6,2%	5,2%	6,9%	14%
Slovenia	16,1%	14,3%	14,9%	25%
Czech Republic	2,4%	4,2%	6,3%	13%
Hungary	2,6%	4,7%	4,3%	13%
Cyprus	2,5%	2,5%	2,9%	13%
EU-25	**7,7%**	**7,5%**	**8,1%**	
Bulgaria	7,1%	9,0%	10,6%	16%
Rumania	13,7%	15,4%	19,2%	24%
EU-27	**7,8%**	**7,7%**	**8,3%**	**20%**

a Target from EU-Guidelines Abstract 2008/0016 (COD)

Source: Energy sector including Grid-losses.
BMU publication „erneuerbare Energien in Zahlen – nationale und international Entwicklung" KI III, Stands July 2008.

2. Potential and Relevance of Photovoltaics for Energy Supply Systems

To reduce greenhouse gas emissions and to ensure the security and sustainability of future energy supply in Europe, a diversification and the involvement of renewable energy sources is necessary. The increasing share of renewable energy is therefore a compulsory ingredient of greenhouse gas mitigation policies, along with a strong requirement to improve energy efficiency. Major policy goals are now recognized as changing living standards and patterns to sustainable ones and ensuring the protection of the environment.

One main reason for issues such as poverty and a variety of health problems is no access to commercial energy, in particular, to an energy grid. Almost a third of the world population does not have access, which seriously hinders development.

In spite of the recent success of renewable energies in some areas, the growing global energy consumption still causes an increase in the consumption of fossil fuels and associated CO_2 emissions. This highlights the urgency to develop and implement renewable energy technologies that can be made available to all people in urban and rural areas and that can make a substantial contribution to the increasing energy demand.

Photovoltaic (PV) is a unique conversation of solar energy in itself. Its assembling of arrays allows a range from a few mill watts up to multi watts installation. PV modules can be part of a consumer product, mounted on roofs of houses, integrated in a building skin or assembled into large power stations. Because of its modularity, it is accepted as a means to serve energy needs in dispersed and isolated communities. It can be designed to be very robust and reliable whilst at the same time it is quiet and safe. PV fits well in the existing infrastructure and it offers possibilities to make intelligent matches between electricity supply and demand.

Reaching the earth's surface total amount of solar energy represents several thousand times the world total energy consumption; the technical potential of converting solar energy directly into heat or electricity is large: greater than 440.000 TWh/year i.e. about four times the earth's total energy consumption.

The main advantage of photovoltaics (PV) as an energy generation technique is the transformation of solar light locally and directly into electricity. PV systems can deliver electric energy to a specific appliance or to an electric grid. It has great potential to play an important role in the transformation towards a sustainable energy supply system of the 21st century to cover a significant share of the energy needs of the world. PV could contribute to the security of future energy supply, provide environmentally benign energy services and enhance economic and social welfare. Alongside other renewable energy technologies and energy efficiency, photovoltaics could become a key technology for the future.

Figure 1: Total PV installed power in selected European countries by the End 2003[6]

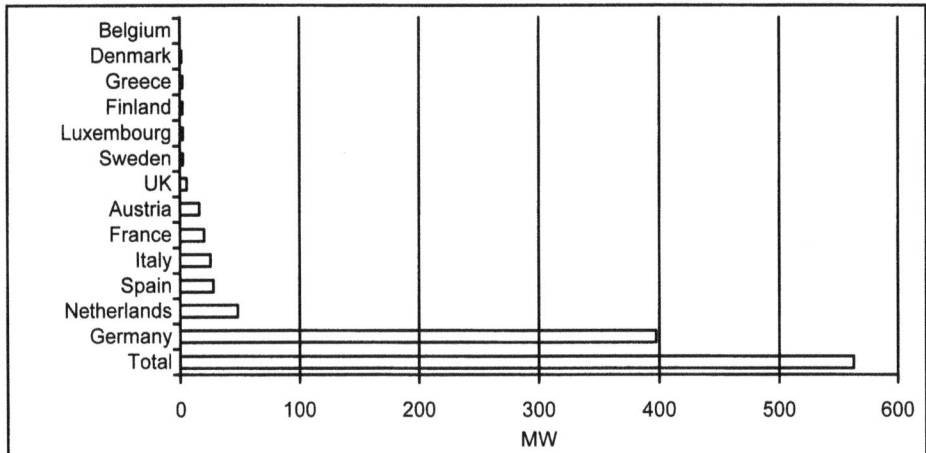

Source: EurObserver 2004

In Europe, fitting the total surface of south-oriented roofs with PV equipment would enable full coverage of our electricity needs. This illustrates that PV could ultimately supply a substantial part or even the majority of our future electricity needs. In view of its excellent technology and market starting position, the EU has a unique opportunity to build a large and highly innovative economic sector, while at the same time developing a key building block for a sustainable energy supply. However, this still requires an ambitious and coherent policy on research and development (R&D), market incentives and communication, and the removal of barriers.

Photovoltaics have always been of less relevance for national energy supply systems, compared with e.g. water or wind, even though PV born electricity has been increased dramatically over the last years. The major barriers preventing uptake in today's market are the costs of PV plants, making the electricity produced too expensive for many applications. Therefore, the present market perception of PV technology is often only a niche application. Figure 1 shows the installed PV power in European countries.

Power production in Europe with Photovoltaic has been mostly accomplished in Germany and Spain. Table 2 below shows the 25 largest solar power plants worldwide.

---

[6] Luxembourg: Office for Official Publications of the European Communities, 2005, ISBN 92-894-8004-1 © European Communities, 2005.

Table 2: The 25 largest Solar power plants worldwide

Power (MWp)	Country	Location	On-grid since
20	Spain	Jumilla	2007
20	Spain	Beneixama	2007
14	USA	Nellis, NV	2007
13,8	Spain	Salamanca	2007
12,7	Spain	Lobosillo	2007
12	Germany	Erlasee / Arnstein	2006
11	Portugal	Serpa	2007
10,35(16,1 MW^3)	Germany	Brandis[1]	2007
10	Germany	Pocking	2006
9,55	Spain	Milagro	2007
8,76	Spain	Viana	2007
8,4	Germany	Göttelborn[2]	2004-2007
8,22	USA	SanLuisVly,Alamosa	2007
6,3	Germany	Mühlhausen	2004
6,277	Spain	Aldea del Conde	2007
6	Spain	Olmedilla	2007
6	Germany	Doberschütz	2007
5,8	Spain	Darro	2007
5,568	Germany	Oberottmarshausen	2007
5,27	Germany	Miegersbach	2005
5,21	Japan	Kameyama	2006
5,076	Germany	Kleinaitingen	2007
5,04	Spain	Alvarado	2007
5	Germany	Thierhaupten	2007
5	Spain	Bullas	2007

[1] *40 MWp Solarpark Waldpolenz is still under construction (10 MWp on-grid since 2007)*
[2] *Plant Göttelborn was constructed in 2004 (4 MWp). 4,4 MWp part added in 2007*
[3] *Construction ongoing – 16,1 MWp in April 2008*

Source: http://www.pvresourcrs.com

Nevertheless, a recent analysis performed by the European PV Technology Research Advisory Council (PV-TRAC)[7] shows that PV has the potential to deliver electricity on a large scale at competitive costs. In 2030 PV could generate 4% electricity worldwide. However, the Council considered 2030 only as an intermediate milestone and stressed that PV would continue to grow steadily well beyond that date. It is envisaged that the technology will develop towards higher

---

[7] Luxembourg: Office for Official Publications of the European Communities, 2005, ISBN 92-894-8004-1 © European Communities, 2005.

efficiency modules, cells and systems, with longer lifetimes and improved reliability, making use of new materials. Generation costs are expected to fall significantly, resulting in increased uptake and deployment both in industrialized markets and for non grid applications in developing countries, thereby creating employment and exports. Ensuring European leadership in this high-tech sector the PV market will require well-coordinated, concentrated and long-term efforts.

2.1 International Market

Shown by the IEA on public expenditure for PV research indicates that Japan invests significantly more public funds in support of PV development than the EU or the US, as shown in Table 3. These figures do not include the cost of feed-in support schemes which are substantial in Japan (PV tariff = 0,3 €/kWh) and in Germany (0,5 €/kWh before 2004).[8]

Table 3: Public Expenditure on PV research and market deployment in 2002

Million US $	R&D	Demonstration	Market Deployment	Total
Japan	59	36	185	280
Europe	58	11	62	131
USA	35	0	80	115
ROW	20	9	13	42
Total	172	56	340	568

Sources: IEA PVPS, annual reports 2002 & 2003, member country contributions.
    Luxembourg: Office for Official Publications of the European Communities, 2005, ISBN 92-894-8004-1 © European Communities, 2005

At the end of 2003 in China there was 50 MW installed capacity of PV, of which 10 MW was installed in 2003. However, the Chinese industry is increasingly visible at international PV fairs and forums. PV seems to be part of the Government's drive toward a strong development of RES; the objective announced at the World Renewable Energy Conference held in Bonn in June 2004 being a 10% share for RES of electricity generation by 2010.

3. Applied Techniques

Governments in Europe have been fostering solar generated power through the last decades. In commerce and science, it is the main challenge to improve and integrate this technology more into industrial and public energy generation. Over

---

[8] Luxembourg: Office for Official Publications of the European Communities, 2005, ISBN 92-894-8004-1 © European Communities, 2005.

the years a variety of different solar cells and Materials have been developed and improved also as a result of silicon shortage reasonable of the emerging demand for silicon through industries, even though silicon is far from rare. In fact, it makes up over a quarter of the Earth's crust, 25,8 weight percent in form of siliceous minerals or silicon dioxide. But it needs to be refined and processed before it can be used in the semiconductor industry; therefore Silicon is a key component also in Information technology such as computer chips. This means that solar companies have to compete with every cell phone, iPod, digital camera and home computer for the silicon available on the market.

Figure 2: Distribution of Cell production by technology[a] and Regional Market share Photovoltaic 2003

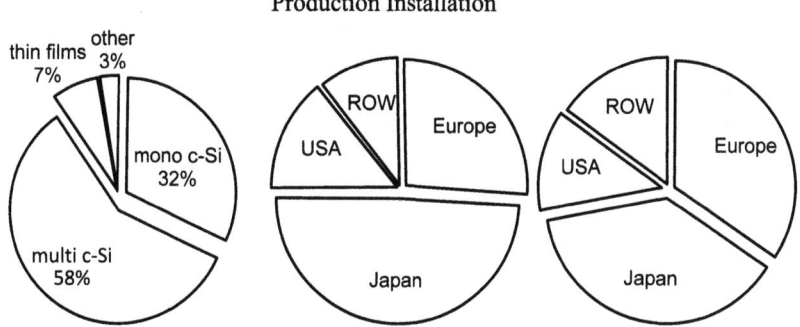

a P.D. Maycock, PV Market update, 2003

Sources: PVNET European Roadmap for R&D, 2004, EUR 21087 EN.
Luxembourg: Office for Official Publications of the European Communities, 2005, ISBN 92-894-8004-1 © European Communities, 2005.

3.1 Commercial application

There are three main groups of base material for solar cells and their production process (s. Table 4):

- wafer based crystalline silicon (c-Si);
- thin films, which include copper-indium / gallium-selenide/sulphide (CIGS), amorphous silicon (a-Si) and cadmium telluride (CdTe) and thin film silicon;
- Nano-structured material to use for coloured cells.

Table 4: Cell-types

Wafer based crystalline cells (c-Si)		Thin film cells	Nanostructured solar cells
Monocrystalline Cells	Polycrystalline cells	Amorphous Silicon cells (a-Si)	Nanostructured CIS solar cells
Standard Silicon p-doped (Cz)	Polycrystalline cells Band drawn silicon, EFG, String Ribbon, APox	Copper-Indium-di-Selenide cells (CIS)	Colored solar cells
High performance silicon cells (Fz) n-doped		CIS Band cells CIS Glass ball cells Copper-Indium-di-Sulfide cells	Polymer organic cells
Ball cells (Spheral cells)		Cadmium-Telluride cells (CdTe)	
Stripe cells (Sliver cells)		Microcrystalline or Micromorph silicon thin film cells (CSG)	
		Concentrator cells (II-V Semiconductor)	
		Hybrid HIT solar cell	

Source: PV3_D4UT_ZellenModule2008

Wafer-based crystalline silicon has a dominant established market position, it has proven reliability and the basic knowledge well developed for the electronics industry. Crystalline Silicon modules are typically produced by growing ingots of silicon in several ways (in a manner similar to that used for the electronics industry), slicing the ingot to make wafers, processing these wafers into solar cells, electrically interconnecting the cells, and encapsulating the strings of cells to form a module. Since high-quality feedstock is required and a large fraction of the silicon is lost during processing, material costs are high compared to thin film modules. In addition, manufacturing is often not yet optimally automated. Finally, current silicon feedstock production is energy intensive. Together with the low silicon utilisation, this leads to a module energy pay-back time of several years, although much shorter than the module lifetime. However crystalline silicon also found its way in the thin film technology and successful research has been made in reducing silicon usage.

Thin-film cells are more in the focus of development and research because of the high energetic production process of Wafer based silicon and a supply shortage of silicon, they are made by coating and patterning entire sheets of substrate, generally glass or stainless steel, with micron-thin layers of conducting and semiconductor materials, followed by encapsulation. This leads to a process that can be highly efficient in materials utilisation, has relatively low labor requirements, and uses comparatively little energy in the total manufacturing process.

3.2 Natural Limits in the Efficiency of PV Cells

Research is also being done to increase the level of efficiency, in addition to optimizing production processes, in order to lower the costs of solar cells. However, different loss mechanisms are setting limits on these plans. Basically, the different semiconductor materials or combinations are suited only for specific spectral ranges. Therefore a specific portion of the radiant energy cannot be used, because the light quanta (photons) do not have enough energy to "activate" the charge carriers. On the other hand, a certain amount of surplus photon energy is transformed into heat rather than into electrical energy. In addition to that, there are optical losses, such as the shadowing of the cell surface through contact with the glass surface or reflection of incoming rays on the cell surface.

Figure 3: Band Gaps in eV

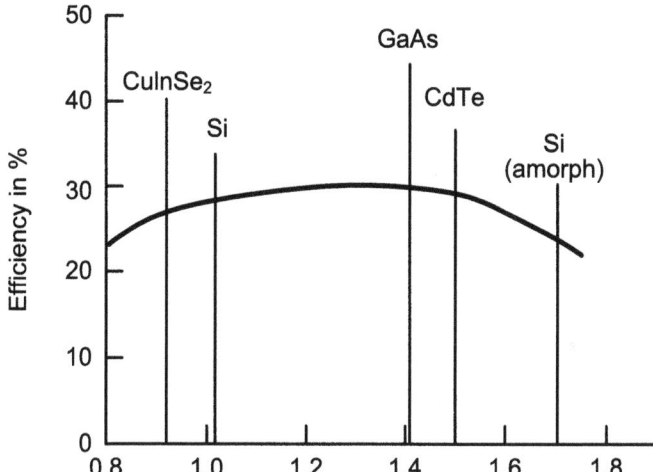

Source: http://solarserver.de

Other loss mechanisms are electrical resistance losses in the semiconductor and the connecting cable. The disrupting influence of material contamination, surface effects and crystal defects, however, are also significant. Single loss mechanisms (photons with too little energy are not absorbed, surplus photon energy is transformed into heat) cannot be further improved because of inherent physical limits imposed by the materials themselves. This leads to a theoretical maximum level of efficiency, i.e. approximately 28% for crystal silicon.

3.3 Wafer-based crystalline silicon solar cells

The semiconductors used for photovoltaic are contaminated or "doped". "Doping" is the intentional introduction of chemical elements, with which one can obtain a surplus of either positive charge carriers (p-conducting semiconductor layer) or negative charge carriers (n-conducting semiconductor layer) from the semiconductor material. If two differently contaminated semiconductor layers are combined, then a so-called p-n-junction results on the boundary of the layers. At this junction, an interior electric field is built up which leads to the separation of the charge carriers that are released by light. Through metal contacts, an electric charge can be tapped. If the outer circuit is closed, meaning a consumer is connected, direct current flows. Silicon cells are approximately 10 cm by 10 cm large (recently also 15 cm by 15 cm). A transparent anti-reflection film protects the cell and decreases reflective loss on the cell surface (s.a. Figure 4).

Figure 4: Wafer based crystalline silicon cells

thickness of the solar cell: approx 0,3 mm
thickness of the n-semiconductor layer: approx 0,002 mm

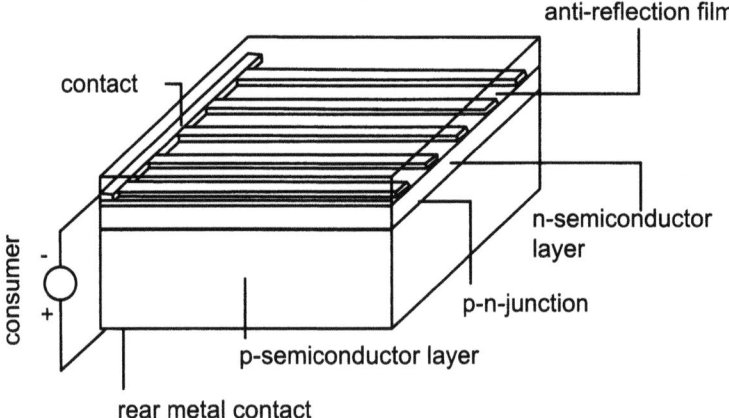

Source: http://solarserver.de

Due to high production costs of crystalline silicon cells and other drawbacks many different variations haven been researched on and have been developed, improved and applied.

3.4 Monocrystalline silicon

Figure 5: Standard silicon p-doped

Source: http://www.svmi.com

Standard silicon p-doped is normally made by the Czochralski (Cz) process. In order to produce monocrystalline ingots by the Czochralski process, high-purity silicon is first loaded into a round quartz crucible and melted. Thereafter, a seed crystal shaped as a thin rod is dipped into the molten silicon. The seed crystal's rod is pulled upwards and rotated at the same time. By precisely controlling the temperature gradients, rate of pulling and speed of rotation, it is possible to extract a large, single-crystal, cylindrical ingot from the melt. This process is normally performed in an inert atmosphere.

Figure 6: High performance silicon cells

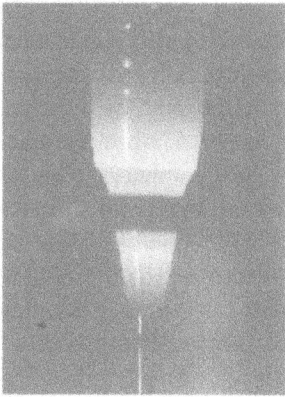

Source: http://www.igafa.de

High performance silicon cells (Fz) crystallization of float zone ingots a high-purity alternative to the Czochralski process. A radio frequency (RF) field is used to produce a local melted zone on the polycrystalline rod, without the liquid being in contact with anything except silicon. The rod is moved relative to the RF field so that the molten (float) zone is moving across the rod. A seed crystal is used at one end in order to start the growth. This molten zone carries the impurities away with it, reducing impurity concentration.

Figure 7: Spheral solar cells ball

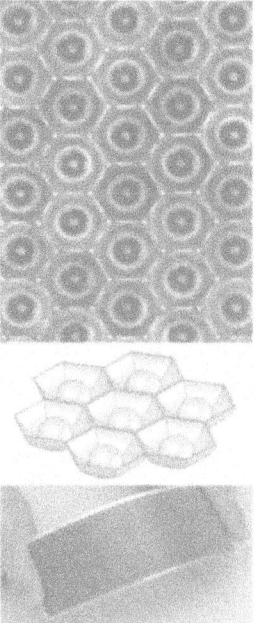

Source: http://www.technologyreview.com

Spheral solar cells ball cells technology used in manufacturing spheral solar cells involves bonding tiny silicon spheres between sheets of thin and flexible substrates (usually aluminium). The front foil acts as the electrically-negative contact and determines the spacing of the spheres, while the back foil acts as the electrically-positively contact to the core of the spheres. The fabrication of the photovoltaic system involves sphere fabrication, sphere junction formation and finishing, cell fabrication (involving setting up of the spheres between substrate sheets) and assembly of the modules. Solar energy is absorbed by the silicon spheres from different angles as the sun moves over different directions, extending the period of conversion of energy, while at the same time using lesser amounts of silicon in the cell manufacturing process.

Figure 8: Sliver solar cells technology

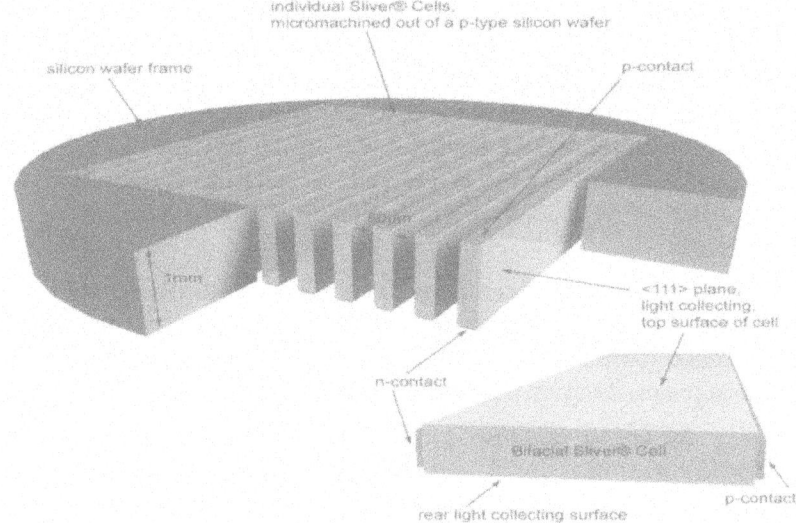

Source: http://spie.org/Images/Graphics/Newsroom/Imported-2009/1593/1593_fig2.jpg

Sliver solar cells technology uses standard materials and conventional techniques in novel ways to create thin single crystalline solar cells with superior performance at significantly reduced cost. Sliver cells are fabricated on single crystal silicon. Sliver modules are manufactured using techniques adapted from conventional module manufacture. Sliver modules can be efficient, low cost, bifacial, transparent, flexible, shadow-tolerant and light-weight. Sliver technology has the potential to be a comprehensive long-term solution for PV. Standard single crystal silicon wafers approximately 1 mm thick are used as the starting material for the sliver cell process. Low cost micromachining methods are used to create many narrow parallel grooves that extend vertically through the wafer but do not extend to the wafer edge. The grooves lead to the creation of an array of thin, parallel, silicon strips, referred to as "Slivers", confined in the wafer, and held in place at their ends by the un-grooved part of the wafer, referred to as the wafer frame. The entire wafer, containing up to several thousand slivers, is then processed using standard techniques to turn each of the slivers into a solar cell. At the end of the process, the slivers are cut out of the wafer frame, laid flat, and electrically connected. The rotation of each sliver through 90 degrees generates a large gain in the active surface area – "area multiplication" – compared with the starting wafer.

3.5 Polycrystalline silicon

Polycrystalline cell material contains many single crystals about the thickness of a human hair, or about 1/1.000 the size of the crystals in monocrystalline material. Polycrystalline wafers are made by a casting process in which molten silicon is poured into a mould and allowed to set. Then it is sliced into wafers. This processing has a material loss of about 40% in the process of sawing body in wafers. As well thickness has to be assured for mechanical stability of 0,3 mm. Therefore different ways have been developed to reduce material loss and increase efficiency.

The *band drawn polycrystalline silicon* has already the necessary wafer thickness and just need to be laser cut in planar even pieces. Research and development are trying to reduce the thickness down to 1mm. This technique has compared to conventional processes great potential of cost reduction.

EFG processed polycrystalline silicon (Edge-defined film-fed Growth) has beenproduced since years in industrial batch production. A graphite carrier is led into molten silicon and pulled out slowly. As a result 6,5m long octagonal tubes with a wall thickness of 0,3mm and side lengths 10cm or 12,5cm. However 8 wafers are produced with around 10% material loss. These wafers are polycrystalline but in its electrical quality and its appearance like mono-crystalline silicon.

Figure 9: EFG processed polycrystalline silicon

Source: http://www.gtsolar.com

String Ribbon Process: Two extreme heated carbon strings are vertically pulledthrough a flat crucible with melted silicon. The liquid silicon forms a skin between the carbon strings and crystallises into an 8 cm ribbon. The production is running continuously, the crucible refilled with melted silicon and the cells rolled up on roll.

Figure 10: String Ribbon Process

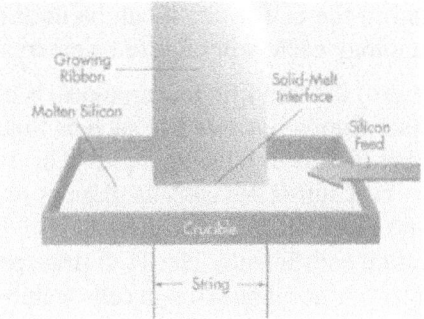

Source: http://www.store.altenergystore.com

The application *wet oxidation for rear surface passivation* significantly reduces the process temperature and therefore prevents the degradation of minority-carrier lifetime. The excellent optical properties of the dielectrically passivated rear surface in combination with plasma textured front surface result in a superior light trapping and allow the use of substrates below 100 μm thickness. A simplified process scheme with laser-fired rear contacts leads to conversion efficiencies of 2 x 3% for multicrystalline and 21 x 2% for monocrystalline silicon solar cells on small device areas (1 cm^2).[9]

3.6 Thin film layer cells

Thin film layer cells experience since the nineties stronger development and research. At this process thin photoactive semiconductors are put on a substrate material, usually glass or even plastic. The procedures to put those layers on are for example, sputtering (cathode evaporation) or through an electrolytic bath. As semiconductors several materials have been developed, such as amorphous silicon, Copper-Indium-di-Selenide (CIS), Cadmium-Telluride cells (CdTe). Due to the high light absorption of those materials, film thickness less then 0,001 mm for the conversion of sunlight is theoretically enough. Materials are more tolerant towards impurities of "foreign" atoms in the semiconductor. In comparison to the manufacturing temperature for crystalline silicon of about 1.500°C, manufacturing temperatures for thin film cells from 200°C to 600°C are required. The high level of automation, reduced energy and material demands are interesting features compared to the crystalline silicon technology.

---

[9] Http://www.ise.fhg.de/veroeffentlichungen/nach-jahrgaengen/2004/multicrystalline-silicon-solar-cells-exceeding-20-efficiency.

Thin film layer cells are producible in any size, not as with standard wafer sizes. The wiring is integrated in the cell itself; it can be used as a design feature for example, in wafer technology each wafer has to be externally soldered together.

Amorphous silicon is one of the thin film technologies, is made by depositing silicon onto a glass substrate from a reactive gas such as Silane (SiH_4). Amorphous silicon is one of a number of thin film technologies. This type of solar cell can be applied as a film to low cost substrates such as glass or plastic. Despite low efficiency thin film layer have capability to use efficient also diffuse light and as well have favorable temperature coefficients. The TCO (transparent conductive oxide) represents the front contact. Amorphous silicon cells feature a degradation process where the efficiency reduces around two to five percent.

Figure 11: Amorphous Silicon

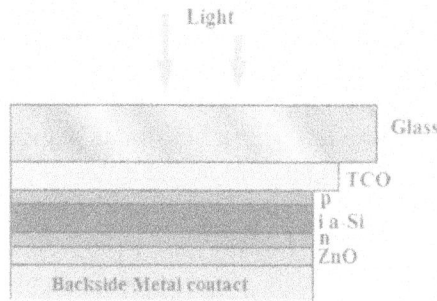

Source: PV3_D4UT_ZellenModule2008

Copper-Indium-di-Selenide is the active semiconductor material in (CIS) thin film technology. Often in CIS cells is additionally Gallium and/or Sulphur added to enhance electrical attributes. CIS cells do not suffer a degradation process such as amorphous silicon cells but need to be protected very well from humidity. Under all thin film technologies these have currently the highest efficiency. Because of insignificant amount of selenium and cadmium CIS cells conform to environmental regulations.

Figure 12: Copper-Indium-di-Selenide

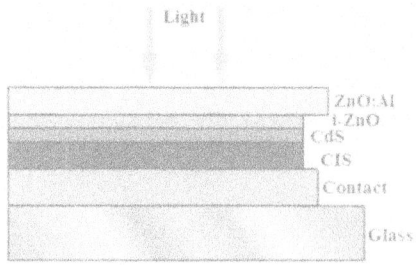

Source: PV3_D4UT_ZellenModule2008

Cadmium-Telluride (CdTe) technique requires least production costs under now a day's thin film technologies. Market acceptance because of Cadmium is not widely given. However, the producers publish environmental impact analyses and a worldwide take-back and recycling system in order to prove the environmental friendliness of the CdTe cells.

Figure 13: Cadmium-Telluride (CdTe)

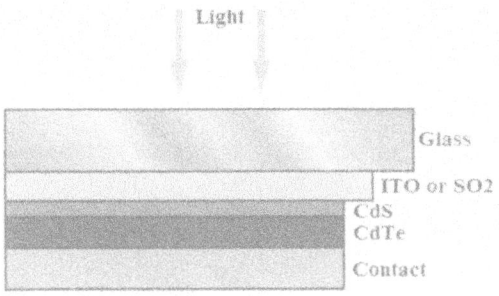

Source: *PV3_D4UT_ZellenModule2008*

The CIS glass cell is a concept where 0,2 mm glass balls represent the solar cells. These are put in thousands on a perforated metal foil and cut out in squares. The light sensitive absorber layer is put onto these glass balls and a metal foil forms the back contact. This concept allows variety of design, which is an important feature in construction building.

Figure 14: The CIS glass cell

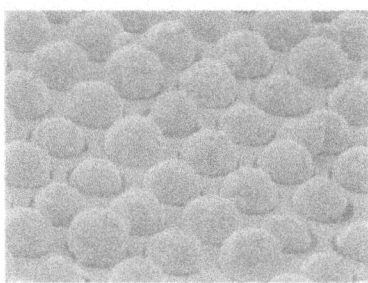

Source: *PV3_D4UT_ZellenModule2008*

3.7 Crystalline silicon thin film layer solar cells

Crystalline silicon thin film layer solar cells are a new direction to use the electrical, ecological material qualities of silicon and the features of thin film layer technique such as high level of automation, reduced energy and material demands.

CSG crystalline silicon on glass has the advantages of amorphous ccrystalline, no heavy metals and in addition no degradation process. The thin polycrystalline layer is 1,5 µm thick, hence less material demand. Amorphous silicon is put on the textured glass substrate and crystallized at 600°C and 900°C. The cell connection occurs through the craters. The back contact is sputtered aluminium. Through direct contact between direct silicon and glass in CSG modules is no polymer interlayer that could hinder sunlight or which qualities could degrade through longer weathering. The Coefficient of Performance (COP) is around 7%.

Figure 15: CSG crystalline silicon on glass

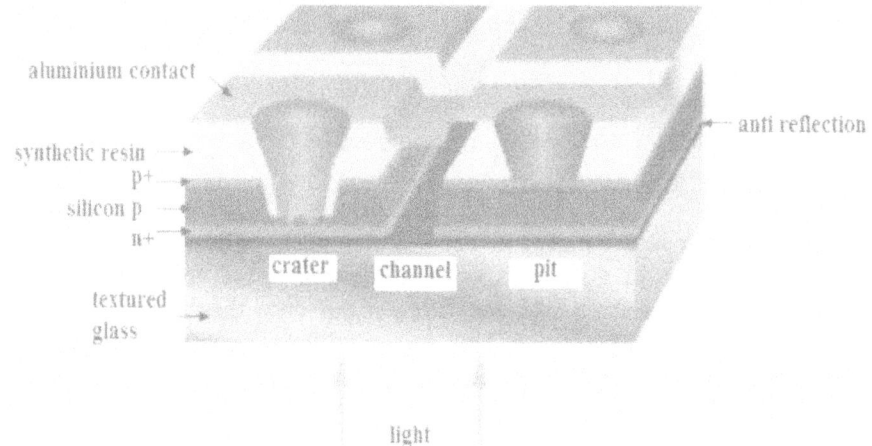

Source: PV3_D4UT_ZellenModule2008

Microcrystalline silicon solar cell technique is with its low temperature deposition based on the use of a classic thin film layer technology such as amorphous silicon solar cells. Because of low process temperatures, costs saving materials for substrates such as plastics are possible. At temperatures between 200°C and 600°C a silicon film with a micro crystalline structure deposits on the substrate. To achieve crystalline silicon layer thickness of 10 µm and less and because of the lower absorbency of crystalline silicon, it is necessary to optimise light incidence with light trapping structures. For that purpose surface of silicon and contact surface is textured with transparent conductive oxide (TCO). Microcrystalline cells have similar optic features as crystalline wafer solar cells.

Figure 16: Microcrystalline silicon solar

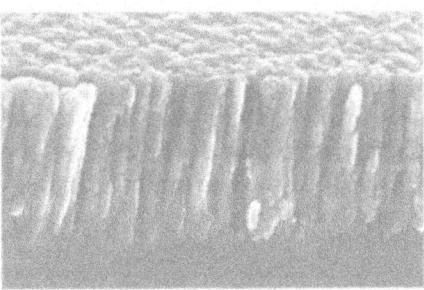

Source: *SEM picture of a microcrystalline Si:H film.Source: Applied Films*

Micromorph silicon solar cell is a type of tandem solar cell. It is a combination between microcrystalline silicon and amorphous silicon. Hereby a 0,3 μm thick layer of amorphous silicon is deposited on glass and at a temperature of 600°C, the surface of the amorphous silicon crystallises in the first 025 μm into microcrystalline silicon. The so resulted tandem cell can advantage more of the sun spectrum, as well in situation with lower radiation the micromoroph silicon solar cell such as amorphous silicon cells, achieves double efficiency and has a lower degradation as amorphous silicon cells.

Figure 17: Micromorph silicon solar cell

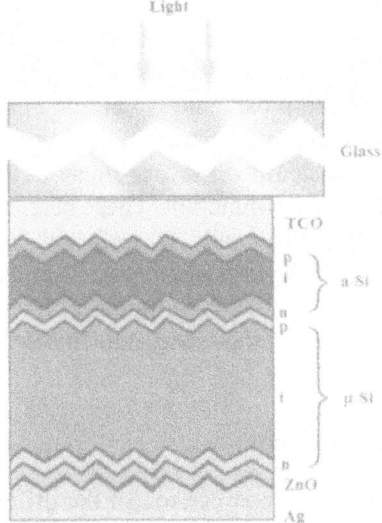

Source: *PV3_D4UT_ZellenModule2008*

The so called *Concentrator cells (II-V Semiconductor)* are combinations of elements of the second and the fifth group of periodic system of elements. For example indium-gallium-arsenide (InGaAs), indium-gallium-phosphide (InGaPh) or germanium can be materials for assembling a high efficient solar cell. Thereby a number of cells are stapled on top of each other, so called multi-junction cell. Light gets concentrated via a lens on a few square millimetres. Current record is held by company "Boeing Spectrolab" with 40,7% efficiency, with a triple-cell GaInP/GaInAs/Ge and 240 times concentrated light. Solar cells with 4 or 5 junctions could reach efficiency to 50%. This technique has found appliance mostly in aerospace or for demonstrating purposes.

Figure 18: Concentrator cells (II-V Semiconductor)

Source: http://www.tms.org

The *heterojunction with intrinsic thin layer* (HIT) cell is a combination of crystalline silicon, amorphous silicon and a third intrinsic thin layer. The core is the monocrystalline silicon wafer placed between amorphous silicon layers. As an interlayer on both sides between crystalline wafer and amorphous silicon, undoped intrinsic (i)-layers of amorphous silicon are placed. A conventional silicon solar cell indicates two differently doped silicon layers where at the HIT cell layers have a different crystal structure a hetero-junction. These cells do not have a degradation process such as amorphous silicon cells, as well with higher temperature performance decreases by 0,33% unlike crystalline silicon at 0,45%. The HIT solar cell uses a wider solar spectrum thus leads to higher efficiency. Deposition temperatures for HIT substrates are 200°C, thus energy and materials demands are lower.

3.8 Comparison of cell qualities

The following table 5 shows the comparison of the different presented cell types. The "Coefficient of Performance" (COP) is the common parameter used to measure the performance of PV cells. "COP (laboratory)" is testing under laboratory conditions: temperature 25°C, 1.000 kW/m^2, 1013hPa ambient pressure; COP (production) is efficiency of in series produced cells and Module COP includes efficiency losses of module assembling, e.g. glass front reflections, wiring connections.

Table 5: Summary of cell qualities

Cell material	COP (laboratory)	COP (production)	Module COP (production)
Concentrator cells (III-V semiconductors)	40,7[a]	27,4	27,0
High performance Monocrystalline silicon	25,0	21,5	20,0
Monocrystalline silicon (Cz)	21,6	17,5	16,0
Hybrid solar cells (HIT)	21,0	18,5	16,8
Polycrystalline silicon	20,3	16,5	15,0
Band drawn silicon	19,7	14,0	13,1
CIS	19,5	11,0	11,0
CdTe	16,5	9,5	10,4
Microcrystalline silicon	15,2	12,0	7,6
CIS Nanocells	14,0	10,5	10,0
Amorphous silicon[b]	13,2	12,5	7,5
Micromorph silicon[b]	13,0	7	11,2
Coloured solar cells	12,0	–	5,0[c]
Organic Solar cells	6,5	–	–

a Measured by concentrated light; b Condition (degradation complete), c Small production
Source: PV3_D4UT_ZellenModule2008

3.9 From solar cell to solar module

Crystalline wafer based silicon cells have cell voltage of about 0,5 Volts and a performance up till 3,75 Watts. To achieve greater voltage, up to 216 cells can be added together to one solar module. The standard solar module nowadays includes mostly 36, 48 or rather 72 cells which are connected in one to four strings in series. Those wiring connection are in the module manufacturing process fully automated. In custom made products wiring might be partly soldered by hand. Thin layer cell modules are fabricated different; the cells are vapor deposited in narrow stripes along the module and within the deposition process, cells are connected in series. Manufacturing process so far is defined by the size of the cells. Crystalline wafer based solar modules have usually four to eight cell rows next to each other. The electrically packaged cell rows are laid packed between glass front and plastic foil whereby the cells will be embedded in Ethylene-Vinyl-Acetate (EVA) for protection against weathering, mechanical stress and humidity. EVA is transparent and isolates cells electrically. Under pressure and heat cells, glass and foil are baked or laminated to a weatherproof and breakproof unit. The front glass is usually a special solar glass hardened and poor of iron oxide for maximum light transmission. The so far cheapest and lightest module is the glass foil module, if the back side is glass as well it is called double glass module. Often module have an aluminum frame to protect fragile glass edges and for installation. Modules without frame are so called laminates. Standard crystalline wafer based silicon modules have about 8 to 20 kg and a plane of 0,6 to 1.5 m^2.

Mostly backside foil is white, grey or blue, the module is opaque (not transparent). In case modules as provided with transparent backside foil, sunlight can shine through between the cells and the module is called semitransparent.

In series connected module need a bypass-diode to prevent overcharge of modules in case of shadowing for example where as in parallel connected cells no further measures need to be done, though here may be higher loss in power under shadowing. In large generation plants both of these wiring techniques are in use to benefit both vantages. Furthermore to promote the most efficiency of the plant, the equipment and the mounting has been improved. The following Table gives an overview of the different cell packaging types.

Table 6: Examples for cell packaging (Ethylene-Vinyl-Acetate (EVA))

EVA packaging	EVA packaging	EVA packaging	EVA packaging
Glass-foil module (c-Si) cells	Double glass module (c-Si) cells	Double glass module Thin layer cells (a-Si)/CdTe	Double glass module Thin layer cells CIS
EVA packaging	EVA packaging	EVA packaging	EVA packaging
Foil module (c-Si) cells	Glass foil module Thin layer cells (a-Si) cells	Metal Foil module (c-Si) cells	Foil module (a-Si) cells

Cast resin packaging	Cast resin packaging	Teflon packaging
Double glass module (c-Si) cells	Triple glass module Thin layer cells	Teflon module (c-Si) cells

Source: PV3_D4UT_ZellenModule2008

3.10 Sun tracking

On annual basis in middle Europe an average of maximum 23% for single axis tracker and 28% for dual axis tracker could be achieved, taking self shading losses into account. An exact forecast could be under consideration local weather data, structure of the sun-field and applied tracking method.

Figure 19: Polar mounting

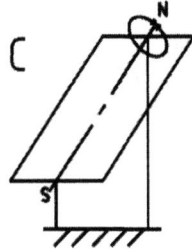

Source: „Ertragspotenzial nachgeführter Photovoltaik in Europa: Anspruch und Wirklichkeit", www.zsw-bw.de

Is the module mounted roughly parallel towards the rotation axis of the earth around the North and South Pole, it is named *polar mounting*. These single axis tracking devices are following diurnal variation and optional yearly variation of the sun. A sun tracker for yearly variation of the sun can be a seasonal manual adjustment.

Figure 20: Horizontal axle sun trackers

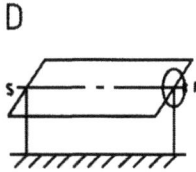

Source: „Ertragspotenzial nachgeführter Photovoltaik in Europa: Anspruch und Wirklichkeit", www.zsw-bw.de

Horizontal axle sun trackers are tracking diurnal variation of the sun. Since these do not tilt toward the equator they are not especially effective during winter mid day, unless the location is near the equator. These horizontal sun trackers are adding a substantial amount of productivity during the spring and summer seasons when the solar path is high in the sky. These devices are less effective at higher latitudes. The principal advantage is the inherent robustness of the supporting structure and the simplicity of the mechanism. Since the panels are horizontal, they accessible for cleaning and maintenance. For active mechanisms, a single control and motor may be used to actuate multiple rows of panels.

Figure 21: Vertical axle

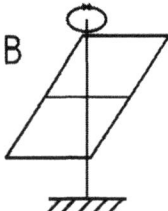

Source: „Ertragspotenzial nachgeführter Photovoltaik in Europa: Anspruch und Wirklichkeit", www.zsw-bw.de

Vertical axle sun trackers may have seasonably adjustable horizontal axis. These trackers are suitable for high latitudes where the solar path is not especially high, but where days in summer are very long and the sun travelling through a long arc.

Figure 22: Altitude, Azimuth sun tracker

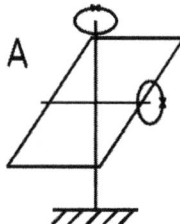

Source: „Ertragspotenzial nachgeführter Photovoltaik in Europa: Anspruch und Wirklichkeit", www.zsw-bw.de

Altitude, Azimuth sun tracker is a mounting system which allows moving in two directions at the same time to locate the specific target. Altitude named is the vertical axis, also weight support and the horizontal axis is named azimuth. However, tracking an object as the earth turns needs to be adjusted on both axes which require a computer.

3.11 Plant Engineering and Energy Conservation

Apart from dimensioning the modules and calculating geographical circumstances aspects of AC/DC converter, by-pass diodes, charge regulator, MPP-Tracker and optional a cooling system for the modules have to been taken into account. This equipment is usually over dimensioned to ensure best performance even in case of bad conditions.

Assembling a standalone system the energy conservation is indispensible. Especially using Photovoltaic with in second's weather changes, clouds etc. can decrease power generation rapidly or increase extremely. There are two types of energy storage over long term and rather short term. Long term is provided with accumulators or short term peak performance well captured with capacitor. In case of feeding the electricity directly into the public grid, the grid takes the role of the conservator.

4. Emissions and consumptions

At the production of solar cells an important factor is manufacturing costs. These differ from coating material and coating material refining. A reason for the high costs of crystalline silicon is the amount of energy involved. Crystalline silicon needs electrolytic refining to ensure its purity and single crystal structure. The energy in use for this process produces causes emissions unless its origin is non fossil.

Cadmium in solar cells: Since the producers designed take-back-systems, recycling strategies and proved the environmental friendliness of these cell types, there seems to be no higher risks in using it.[10]

The production of a photovoltaic module emits approximately 1.600 kg CO_2. This emission could be compensated by the production and substitution of approximately 2.500 kWh electricity based on fossil fuel. The average production of a module, e.g. in Hamburg, Germany is around 825 kWh/kW_p per year.[11]

5. Techniques to consider in the determination of Best Available Techniques

Photovoltaics applicability changes with local conditions and specific implementation. Furthermore the base material and its characteristic are important factors taking into account. Doing this, the following techniques are considered as Best Available Techniques.

5.1 Thin film technologies and several polycrystalline technologies

In fact thin film technologies increase their efficiency and decrease their production costs through national and international research and development. These modules open as well creative opportunities for integrated building service engineering (flexible modules, semitransparent modules, etc.).

However, in the consideration BAT the main aspect is the radiation characteristic of the focused region. At high operating temperatures solar cells generally decrease in their performance. This is stated in the temperature coefficient in mV/K; or in%/K for example Watts per Kelvin. For example the temperature coefficient of HIT cells is round about -0,3%/K compared to crystalline silicon cells with a higher negative influence of increasing ambient temperature of -0,4%/K.[12]

Therefore, choosing the best available photovoltaic technology for Turkey, it should been taken into account, that it is a geographical region with high radiation (see figure 23 and 24) and with high average and maximum temperatures.

---

[10] PV3_D4UT_ZellenModule2008.
[11] Http://wohnen.pege.org/2007-intersolar/photovoltaik-co2.htm.
[12] Http://www.photon.de.

Figure 23: Worlds Radiation

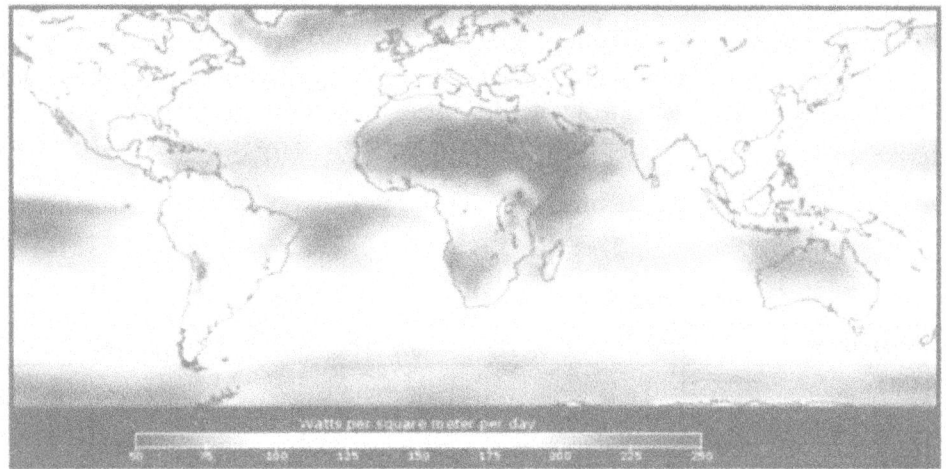

Source: http://www.pvresourcrs.com

Figure 24: Radiation in Europe

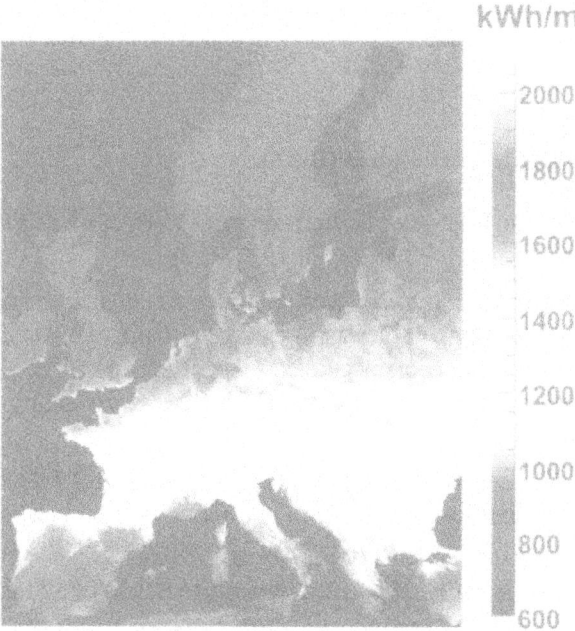

Source: Photon international 03/2006

5.2 Calculation for loss of performance through heat

On standard operating conditions (25°C, 1.013 hPa, 1.000 W/m²) the maximum power point and its temperature dependence can be followed with the temperature coefficients [%/K] illustrated. The difference between operating and standard temperature times the temperature coefficient will show the maximal performance reduction.

$$\left(\vartheta_{op} - \vartheta_{st}\right) \cdot \gamma_{cell} = P_{max} \text{ reduction} \quad (1)$$

Table 7: Effect of Temperature

Modultype	P_{max} reduction [%]	$T_{standard}$ [°C]	$T_{operation}$ [°C]	γ [%/K]
CdTe	-9,2	25	65	-0,23
Amorphous silicon	-9,6	25	65	-0,24
HIT	-12,4	25	65	-0,31
Monocrystalline silicon	-16,4	25	65	-0,41
CIS	-17,6	25	65	-0,44
Polycrystalline silicon	-17,6	25	65	-0,44

Source: http://www.photon.de

6. Best Available Techniques (BAT)

Termination of the Best Available Techniques (BAT) has to be split into two categories:

Areas with high solar radiation such as South Europe: In these regions over the year a high middle temperature is a kind of "standard" condition. Considering this the temperature coefficient takes an important role. Therefore, solar Photovoltaic modules with a low temperature coefficient are more suitable in hot regions over the globe.

For example, CdTe modules with their relatively high COP for thin layer cells have in terms of their low temperature coefficient good qualities.

Areas with lower solar radiation such as North Europe: Crystalline Silicon modules are more sensitive against temperature but have a high COP.

In North Europe lower middle temperatures make the environment more suitable for crystalline modules. Amorphous Silicon modules are suitable as well for Northern Europe or in cloudy region. Thin Layer Amorphous Silicon works also well with spread sunlight such as on cloudy days.

7. Emerging Techniques

Research and development is going into a thin layer crystalline technique. Directions are reducing manufacturing energy and material demand of silicon based cells and improvement and development in the polymer solar cell technology, such as colored and organic solar cells.

Thin layer crystalline silicon cell is an upcoming technique taking benefits from crystalline silicon and thin layer technology. Either cutting at temperatures of about 1.100°C thin film of crystalline silicon and placing it on a cheap wafer equivalent like graphite for example or in a new process where 1.000°C hot trichloridesilane gas leaves a silicon deposition of 20-30 µm thickness on a monocrystalline wafer. After completing the thin layer cell on the wafer, similar to a foil on the wafer, it can be removed and finished. This can be repeated with one monocrystalline wafer about nine times[13].

Nano-structured solar cells are more in the picture of the future including as well colored solar cells or organic cells. Breakthrough has been from the Scientist Alan Heeger in the Year 2000 for the organic photovoltaic effect in itself and received the Nobel prize for Chemistry; With the Korean Kwanghee Lee Alan Heeger presented a organic solar cell with the still unbroken world record of COP 6,5%.[14]

Flexible solar cells[15] such as spheral cells are also in development with other material and techniques, such as nano wires for example. In Ontario at the Mc Master University scientist developed light absorbing nano wires made of photo active material. Similar nano wires could be put on reusable substrates. The little particles could be integrated also into a polyester coat. Both of these appendages could lead to far more flexible and cheaper thin film layer techniques.

---

[13] PV3_D4UT_ZellenModule2008.
[14] PV3_D4UT_ZellenModule2008.
[15] Http://www.heise.de.

8. Glossary

AC/DC converter	Conversion of alternating current to direct current.
COP	The coefficient of performance shows efficiency of technique to transform energy of radiation into electric energy.
MPP Tracker	This unit tracks the *point* of *maximum power*. This point is dependent on irradiance and cell type. The tracker regulates current and voltage of consumer so that operating point is provided at the maximum power point.
By-pass diodes	
(c-Si)	crystalline silicon
(a-Si)	amorphous silicon
Cz	Czochralski process to produce monocrystalline silicon with high puridity

9. References

BMU Publication „Erneuerbare Energien in Zahlen – nationale und international Entwicklung" KI III, Stands July 2008

„Erneuerbare Energien und Klimaschutz" ISBN 978-3-446-41444-0

European Union in Directive 96/61/EC of 24 September 1996 concerning integrated pollution prevention and control Electricity production from renewable sources, technical potential of RES and electricity generation costs, Office for Official Publications of the European Communities, European Communities, 2005

http://eippcb.jrc.ec.europa.eu/

http://„Ertragspotenzial nachgeführter Photovoltaik in Europa: Anspruch und Wirklichkeit" www.zsw-bw.de

http://Photon international 03/2006

http://SEM picture of a microcrystalline Si:H film.Source: Applied Films

http://Solarserver.de

http://wohnen.pege.org/2007-intersolar/photovoltaik-co2.htm

http://www.dgs-berlin.de/fileadmin/PDF/PV3_D4UT.pdf

http://www.Europa.eu/scadplus/leg/en/lvb/l28045.htm

http://www.gtsolar.com

http://www.heise.de

http://www.igafa.de

http://www.ise.fhg.de/veroeffentlichungen/nach-jahrgaengen/2004/multi-crystalline-silicon-solar-cells-exceeding-20-efficiency

http://www.photon.de

http://www.pvresourcrs.com

http://www.store.altenergystore.com

http://www.svmi.com

http://www.technologyreview.com

http://www.tms.org

Luxembourg: Office for Official Publications of the European Communities, 2005 ISBN 92-894-8004-1 © European Communities, 2005

Photon international 03/2006

Reference Document on Best Available Techniques for Management of Tailings and Waste – Rock in Mining Activities (July 2004), Directorate – General JRC Institute for Prospective Technological Studies, sustainability in Industry, Energy and Transport, European IPPC Bureau of the European Commission.

Chapter 6

Best Available Techniques – Working Paper – Waste to Energy (W2E) – Waste incineration –

Kerstin Kuchta, Konstantin Haker

The structure of this document is also derived from an original reference document on Best Available Techniques (BAT). The determination of BAT is used as a tool to display the state-of-the-art of energy technologies with the generally accepted guidelines of the IPCC Directive 2008(1/EC).

Scope

The scope of this document is the description of different techniques for waste incineration plants that are successfully used in practice and which have proved value in industry. In general this paper gives an overview of combustion and flue gas treatment techniques. The incineration of sewage sludge is only mentioned sketchily.

1. General Information

1.1 The purpose and objective of waste incineration

Incineration is used as a treatment for a very wide range of wastes. Incineration itself is commonly only one part of a complex waste treatment system that altogether, provides for the overall management of the broad range of wastes that arise in society. The incineration sector has undergone rapid technological development over the last 10 to 15 years. Much of these changes have been driven by legislation specific to the industry and this has, in particular, reduced emissions to air from individual installations. Continual process development is ongoing, with the sector now developing techniques which limit costs, whilst maintaining or improving environmental performance. The objective of waste incineration, in common with most waste treatments, is to treat waste so as to reduce its volume and hazard, whilst capturing (and thus concentrating) or destroying potentially harmful substances. Incineration processes can also provide a means to enable recovery of the energy, mineral and/or chemical content from waste (IPPC 2006).

1.2 Waste Incineration

Basically, waste incineration (WI) is the oxidation of the combustible materials contained in the waste. Waste is generally a highly heterogeneous material, consisting essentially of organic substances, minerals, metals and water. During incineration, flue-gases are created that will contain the majority of the available fuel energy as heat. The organic substances in the waste will burn when they have reached the necessary ignition temperature and come into contact with oxygen. The actual combustion process takes place in the gas phase in fractions of seconds and simultaneously releases energy. Where the calorific value of the waste and oxygen supply is sufficient, this can lead to a thermal chain reaction and self-supporting combustion, i.e. there is no need for the addition of other fuels. Although approaches vary greatly, the incineration sector may approximately be divided into the following main sub-sectors:

- *Mixed municipal waste incineration* – treating typically mixed and largely untreated household and domestic wastes but may sometimes including certain industrial and commercial wastes (industrial and commercial wastes are also separately incinerated in dedicated industrial or commercial non-hazardous waste incinerators);
- *pretreated municipal or other pretreated waste incineration* – installations that treat wastes that have been selectively collected, pretreated, or prepared in some way, such that the characteristics of the waste differ from mixed waste. *Specifically prepared refuse derived fuel incinerators fall in this subsector;*
- *hazardous waste incineration* – this includes incineration on industrial sites and incineration at merchant plants (that usually receive a very wide variety of wastes) iv. Sewage sludge incineration – in some locations sewage sludges are incinerated separately from other wastes in dedicated installations, in others such waste is combined with other wastes (e.g. municipal wastes) for its incineration;
- *clinical waste incineration* – dedicated installations for the treatment of clinical wastes, typically those arising at hospitals and other healthcare institutions, exist as centralised facilities or on the site of individual hospital etc. In some cases certain clinical wastes are treated in other installations, for example with mixed municipal or hazardous wastes. Data in this document shows that, at the time of its compilation:

Around 20-25% of the municipal solid waste (MSW) produced in the EU-15 is treated by incineration (total MSW production is close to 200 million tonnes per year). The percentage of MSW treated by incineration in individual Member

States of the EU-15 varies from 0% to 62% where the total number of MSW installations in the EU-15 is over 400.

Annual MSW incineration capacity in individual European countries varies from 0 kg to over 550 kg per capita. In Europe the average MSW incinerator capacity is just under 200.000 tonnes per year. The average throughput capacity of the MSWI installations in each Member State (MS) also varies. The smallest plant size average seen is 60.000 tonnes per year and the largest close to 500.000 tonnes per year. Around 12% of the hazardous waste produced in EU-15 is incinerated (total production close to 22 million tonnes per year).

The expansion of the MSW incineration sector is anticipated in Europe over the next 10-15 years as alternatives are sought for the management of wastes diverted from landfill by the Landfill Directive and both existing and new MS examine and implement their waste management strategies in the light of this legislation.

1.3 Waste Incineration in Turkey

The most frequent method for the disposal of wastes in Turkey is the irregular landfilling and the methods of regular landfilling. Composting, incineration or recycling are not common. According to the data obtained from Turkish Statistical Institute; approximately 40% of hazardous wastes generated is recycled (including incineration). A certain part of hazardous wastes produced by industry is recycled on site by the industrial facilities themselves.

İzmit Waste and Purifying Wastes, Incineration and Valuation Inc. (İZAYDAŞ) established in 1996 by İzmit Metropolitan Municipality is the only Hazardous Waste Incineration Facility of Turkey. The facility, which has an incineration capacity of 35.000 tons and 65.000 tons storage capacity, can dispose 5% of 2 million tons of hazardous wastes produced annually.

According to Environment Situation Report of 2005 prepared by İstanbul Provincial Directorate of the Ministry of Environment and Forestry; only 7.763 tons (1%) of the 750.000 tons of hazardous wastes produced annually in İstanbul is sent to İZAYDAŞ, the sole facility to which this type of wastes can be sent. The remaining part is either used in reproduction or disposed to landfills with household wastes. This, in turn, creates a serious threat to environment and human health through contaminating underground waters, agricultural production or direct contact.

A second waste incineration facility was established in Menemen (İzmir). Because of non adequate waste material in this region this facility cannot be operated.

"Investment Plan Specific to Directive" prepared for harmonizing with EU Acquis; the total amount of hazardous wastes is envisaged to be approximately

1.060.000 tons (approximately 650.000 tons for landfilling and 410.000 tons for incineration). These data show clearly, how insufficient the existing capacity in Turkey is and how high the potentials for improvements in the Turkish waste management sector, especially in waste incineration sector are (TCA 2007).

2. Applied Techniques

Only an overview of technical equipment that is implemented in waste incineration plants can be given in this paper. Every incineration plant has its own and special waste processing system so that a resilient conclusion concerning BAT in waste progression procedures is impossible. Therefore applied techniques are listed and briefly described in the following.

Figure 1: Layout of a Municipal solid waste incineration plant

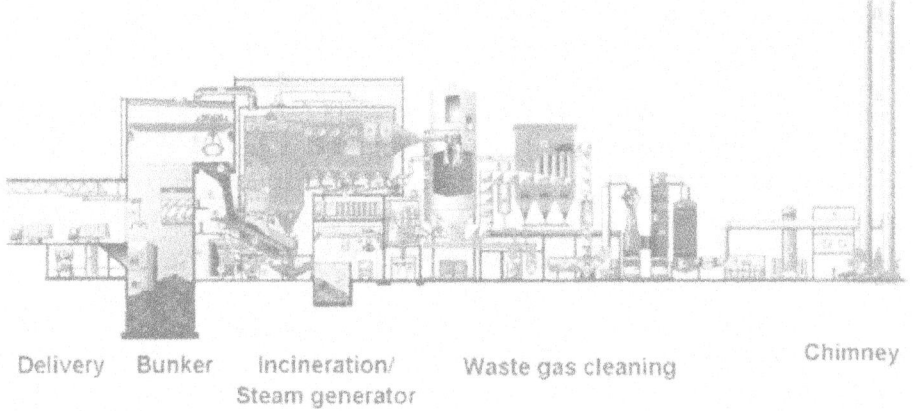

Source: IPPC 2006

2.1 Common techniques applied in the sector of waste incineration

Most waste incineration and co-incineration plants, are setup in 5 different main parts (see Figure 1). Incoming material is brought to the delivery sector which is connected to the waste bunker where the material is dropped off. In a following step the waste is optionally pre-treated (depending on the type of waste and incinerator) and then processed to the incineration chamber. Produced heat is used for the generation of water steam which runs turbines for the production of electricity. The flue gas or waste gas is transported to the waste/fuel gas cleaning section which is a big part of an incineration plant. There pollutants and contraries are removed from the gas so that the offgas fulfils the thresholds required by legal regulations that belong to the incineration plant.

BAT – Working Paper – Waste to Energy (W2E) – Waste incineration 151

Although incineration is by far the most widely applied, there are three main types of thermal waste treatment relevant to this working paper:

- pyrolysis – the thermal degradation of organic material in the absence of oxygen;
- gasification – the partial oxidation;
- incineration – the full oxidative combustion.

The reaction conditions for these thermal treatments vary, but may be differentiated approximately as follows.

Table 1: Typical reaction conditions and products from pyrolysis, gasification and incineration processes

	Pyrolysis	Gasification	Combustion
Reaction temperature [°C]	250-700	500-1.600	800-1.450
Pressure (bar)	1	1-45	1
Atmosphere	Inert/nitrogen	Gasification agent: O_2, H_2O	Air
Stoichiometric ratio	0	<1	>1
Products from the process			
Gas phase:	H_2, CO, hydrocarbons, H_2O, N_2	H_2, CO, CO_2, CH_4, H_2O, N_2	CO_2, H_2O, O_2, N_2
Solid phase:	Ash, coke	Slag, ash	Ash, slag
Liquid phase:	Pyrolysis oil and water		

Source: IPPC 2006

2.2 The thermal treatment stage

Different types of thermal treatments are applied to the different types of wastes, however not all thermal treatments are suited to all wastes. The concepts and applications behind the most common technologies, in particular are:

- Grate incinerators;
- rotary kilns;
- fluidised beds;
- pyrolysis and gasification systems.

Some other, more specific technologies are:

- Municipal solid waste;
- incineration of sewage sludge;
- incineration of hazardous and medical waste.

Municipal solid waste – can be incinerated in several combustion systems including travelling grate, rotary kilns, and fluidised beds. Fluidised bed technology requires MSW to be of a certain particle size range – this usually requires some degree of pretreatment and/or the selective collection of the waste.

Incineration of sewage sludge – this takes place in rotary kilns, multiple hearth, or fluidised bed incinerators. Co-combustion in grate-firing systems, coal combustion plants and industrial processes is also applied. Sewage sludge often has a high water content and therefore usually requires drying, or the addition of supplementary fuels to ensure stable and efficient combustion.

Incineration of hazardous and medical waste – rotary kilns are most commonly used, but grate incinerators (including co-firing with other wastes) are also sometimes applied to solid wastes, and fluidised bed incinerators to some pretreated materials. Static furnaces are also widely applied at on-site facilities at chemical plants.

The following table 2 shows the single techniques and their application status ordered by the main waste types.

Table 2: Summary of the current successful applications to the main waste types

Technique	Untreated Municipal waste	Pretreated MSW and RDF	Hazardous waste	Sewage sludge	Clinical waste
Grate – reciprocating	Widely applied	Widely Applied	Not normally applied	Not normally applied	Applied
Grate – travelling	Applied	Applied	Rarely applied	Not normally applied	Applied
Grate – rocking	Applied	Applied	Rarely applied	Not normally applied	Applied
Grate – roller	Applied	Widely Applied	Rarely applied	Not normally applied	Applied
Grate – water cooled	Applied	Applied	Rarely applied	Not normally applied	Applied
Grate plus rotary kiln	Applied	Not normally applied	Rarely applied	Not normally applied	Applied
Rotary kiln	Not normally applied	Applied	Widely applied	Applied	Widely applied

Technique	Untreated Municipal waste	Pretreated MSW and RDF	Hazardous waste	Sewage sludge	Clinical waste
Rotary kiln – water cooled	Not normally applied	Applied	Applied	Applied	Applied
Static hearth	Not normally applied	Not normally applied	Applied	Not normally applied	Widely applied
Static furnace	Not normally applied	Not normally applied	Widely applied	Not normally applied	Applied
Fluid bed – bubbling	Rarely applied	Applied	Not normally applied	Applied	Not normally applied
Fluid bed – circulating	Rarely applied	Applied	Not normally applied	Widely applied	Not normally applied
Fluid bed – rotating	Applied	Applied	Not normally applied	Applied	Applied
Pyrolysis	Rarely applied	Rarely applied	Rarely applied	Rarely applied	Rarely applied
Gasification	Rarely applied	Rarely applied	Rarely applied	Rarely applied	Rarely applied

Note: This table only considers the application of the technologies described at dedicated installations. It does not therefore include detailed consideration of the situations where more than one type of waste is processed

Source: IPC 2006

Other processes have been developed that are based on the de-coupling of the phases which also take place in an incinerator: drying, volatilisation, pyrolysis, carbonisation and oxidation of the waste. Gasification using gasifying agents such as, steam, air, carbon-oxides or oxygen is also applied. These processes aim to reduce flue-gas volumes and associated flue-gas treatment costs. Some of these developments met technical and economical problems when they were scaled-up to commercial, industrial sizes, and are therefore pursued no longer. Some are used on a commercial basis (e.g. in Japan) and others are being tested in demonstration plants throughout Europe, but still have only a small share of the overall treatment capacity when compared to incineration (IPPC 2006).

2.2.1 Grate Incinerators

Grate incinerators are widely applied for the incineration of mixed municipal wastes. In Europe approximately 90% of installations treating MSW use grates. Other wastes commonly treated in grate incinerators, often in addition to MSW, include commercial and industrial non-hazardous wastes, sewage sludges and certain clinical wastes.

The waste is discharged from the storage bunker into the feeding chute by an overhead crane, and then fed into the grate system by a hydraulic ramp or another conveying system. The grate moves the waste through the various zones of the combustion chamber in a tumbling motion. The material must survive occasional hopper fires unscathed. The waste hopper may sometimes be fed by a conveyor. In that case, the overhead crane discharges waste into an intermediate hopper that feeds the conveyor (IPPC 2006).

If the delivered waste has not been pretreated, it is generally very heterogeneous in both size and nature. The feed hopper is therefore dimensioned in such a way that bulky materials fall through and bridge formations and blockages are avoided. These blockages must be avoided as they can result in uneven feeding to the furnace and uncontrolled air ingress to the furnace. Feeder chute walls can be protected from heat using:

- Water-cooled double shell construction;
- membrane wall construction;
- water-cooled stop valves;
- fireproof brick lining.

If the feed chute is empty, stop valve equipment (e.g. door seals) can be used to avoid flashbacks and for the prevention of uncontrolled air infiltration into the furnaces. A uniform amount of waste in the filling chute is recommended for uniform furnace management.

Figure 2: Grate, furnace and heat recovery stages of an example MSW incineration plant

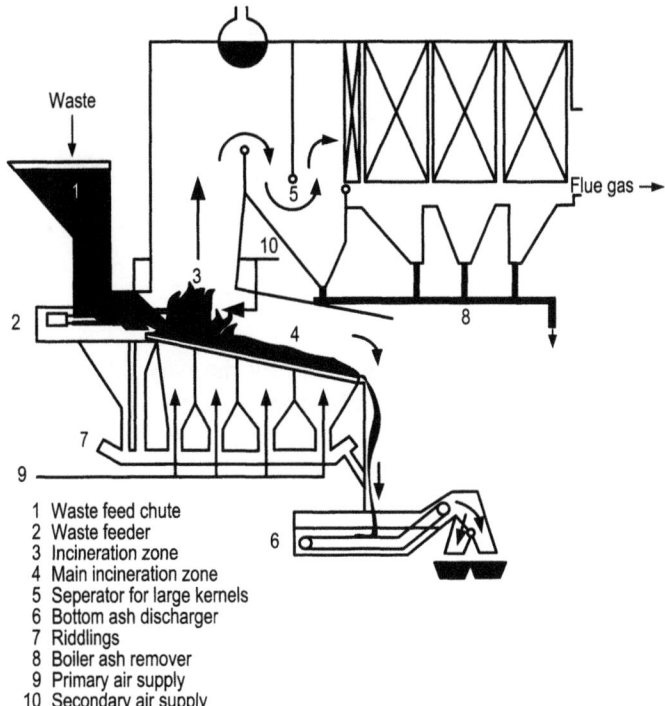

1 Waste feed chute
2 Waste feeder
3 Incineration zone
4 Main incineration zone
5 Seperator for large kernels
6 Bottom ash discharger
7 Riddlings
8 Boiler ash remover
9 Primary air supply
10 Secondary air supply

Source: IPPC 2006

A target of the incineration grate is a good distribution of the incineration air into the furnace, according to combustion requirements. A primary air blower forces incineration air through small grate layer openings into the fuel layer. More air is generally added above the waste bed to complete combustion. It is common for some fine material (sometimes called *riddlings or siftings*) to fall through the grate. This material is recovered in the bottom ash remover. Sometimes it is recovered separately and may be recycled to the grate for repeated incineration or removed directly for disposal. When the sifting is recirculated in the hopper, care should be taken not to ignite the waste in the hopper. Normally, the residence time of the wastes on the grates is not more than 60 minutes.

In general, one can differentiate between continuous (roller and chain grates) and discontinuous feeder principles (push grates). Figure 3 shows some types of grates:

Figure 3: Different grate types

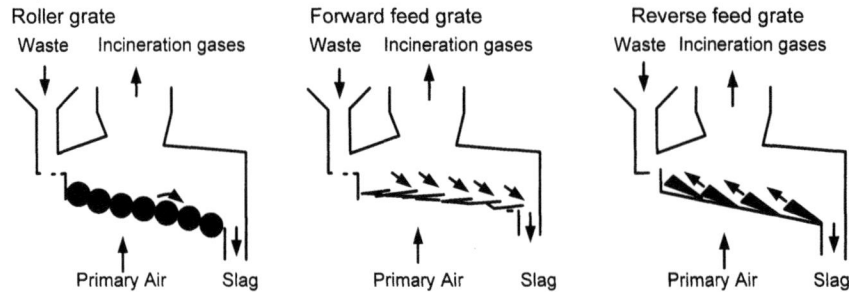

Source: IPPC 2006

Different grate systems can be distinguished by the way the waste is conveyed through the different zones in the combustion chamber. Each has to fulfill requirements regarding primary air feeding, conveying velocity and raking, as well as mixing of the waste. Other features may include additional controls, or a more robust construction to withstand the severe conditions in the combustion chamber.

2.2.2 Rotary kiln

Figure 4: Example of a rotary kiln

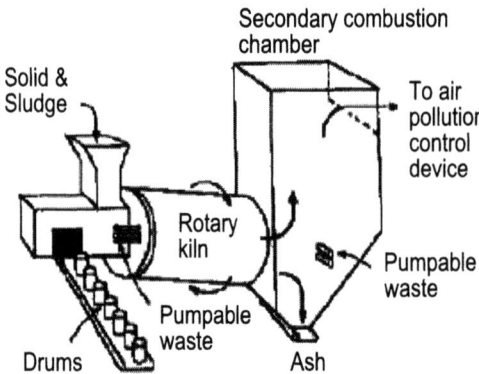

Source: IPPC 2006

Rotary kilns are very robust and almost any waste, regardless of type and composition, can be incinerated. Rotary kilns are, in particular, very widely applied for the incineration of hazardous wastes. The technology is also commonly used for clinical wastes (most hazardous clinical waste is incinerated in high temperature rotary kiln incinerators, but less so for municipal wastes. Operating temperatures of rotary kilns used for wastes range from around 500°C (as a gasifier) to 1.450°C (as a high temperature ash melting kiln). Higher temperatures are sometimes encountered, but usually in non-waste applications. When used for conventional oxidative combustion, the temperature is generally above 850°C. Temperatures in the range 900-1.200°C are typical when incinerating hazardous wastes. Generally, and depending on the waste input, the higher the operating temperature, the greater the risk of fouling and thermal stress damage to the refractory kiln lining. Some kilns have a cooling jacket (using air or water) that helps to extend refractory life, and therefore the time between maintenance shut-downs.

2.2.3 Fluidised Beds

Fluidised bed incinerators are widely applied to the incineration of finely divided wastes e.g. RDF and sewage sludge. It has been used for decades, mainly for the combustion of homogeneous fuels. Among these are coal, raw lignite, sewage sludge, and biomass (e.g. wood). The fluidised bed incinerator is a lined combustion chamber in the form of a vertical cylinder. In the lower section, a bed of inert material, (e.g., sand or ash) on a grate or distribution plate is fluidised with air. The waste for incineration is continuously fed into the fluidised sand bed from the top or side.

Preheated air is introduced into the combustion chamber via openings in the bed-plate, forming a fluidised bed with the sand contained in the combustion chamber. The waste is fed to the reactor via a pump, a star feeder or a screw-tube conveyor. In the fluidised bed, drying, volatilisation, ignition, and combustion take place. The temperature in the free space above the bed (the freeboard) is generally between 850 and 950°C. Above the fluidised bed material, the free board is designed to allow retention of the gases in a combustion zone. In the bed itself the temperature is lower (around 650°C).

Figure 5: Different types of fluidised bed incinerators

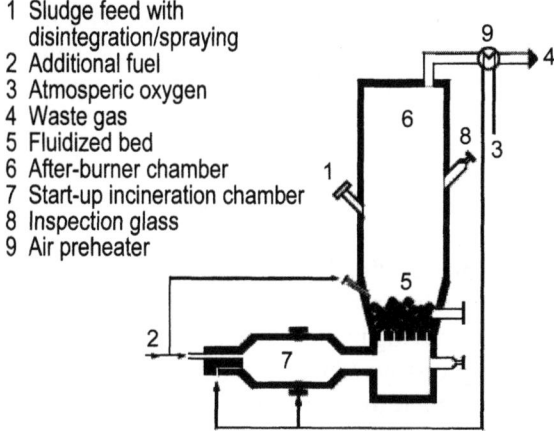

1 Sludge feed with disintegration/spraying
2 Additional fuel
3 Atmosperic oxygen
4 Waste gas
5 Fluidized bed
6 After-burner chamber
7 Start-up incineration chamber
8 Inspection glass
9 Air preheater

Source: IPPC 2006

This type of fluidised bed is commonly used for sewage sludge, as well as for other industrial sludges e.g. petrochemical and chemical industry sludges. The stationary, or bubbling fluidised bed, consists of a cylindrical or rectangular lined incineration chamber, a nozzle bed, and a start-up burner located below.

The waste can be loaded via the head, on the sides with belt-charging machines, or directly injected into the fluidised bed. In the bed, the waste is crushed and mixed with hot bed material, dried and partially incinerated. The remaining fractions (volatile and fine particles) are incinerated above the fluidised bed in the freeboard. The remaining ash is removed with the flue-gas at the head of the furnace.

Figure 6: Different types of fluidised bed incinerators

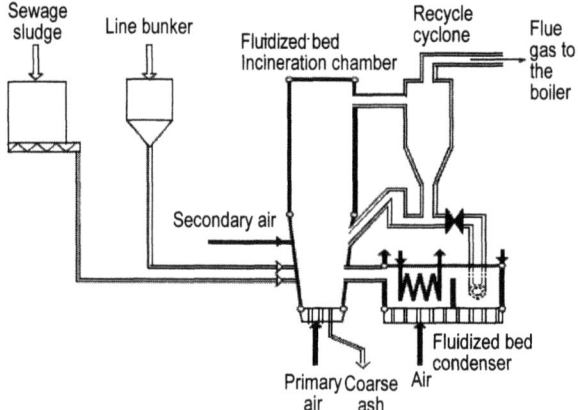

Source: IPPC 2006

The circulating fluidised bed is especially appropriate for the incineration of dried sewage sludge with a high heat value. It works with fine bed material and at high gas speeds that remove the greater part of the solid material particles from the fluidised bed chamber with the flue-gas. The particles are then separated in a downstream cyclone and returned to the incineration chamber and is incinerated there at 850-950°C.

The *spreader stoker furnace system* may be considered as an intermediate system between grate and fluidised bed incineration. The waste (e.g. RDF, sludge etc) is blown into the furnace pneumatically at a height of several meters. Fine particles participate directly in the incineration process, while the larger particles fall on the travelling grate, which is moving in the opposite direction to the waste injection. As the largest particles are spread over the greatest distance, they spend the longest time on the grate in order to complete the incineration process. Secondary air is injected to ensure that the flue-gases are adequately mixed in the incineration zone. Compared to grate incineration the grate is of less complicated construction due to the relatively smaller thermal and mechanical load. When compared to fluidised bed systems the uniformity of particle size is less important and that there is a lower risk of clogging.

The *Rotating fluidised beds* is a development of bubbling bed for waste incineration. Inclined nozzle plates, wide bed ash extraction chutes and upsized feeding and extraction screws are specific features to ensure reliable handling of solid waste. Temperature control within the refractory lined combustion chamber (bed and freeboard) is by flue-gas recirculation. This allows a wide range of calorific value of fuels, e.g. co-combustion of sludges and pretreated wastes.

2.2.4 Pyrolysis and Gasification

Alternative technologies for thermal waste treatment have been developed since the 1970s. In general these have been applied to selected waste streams and on a smaller scale than incineration. These technologies attempt to separate the components of the reactions that occur in conventional waste incineration plants by controlling process temperatures and pressures in specially designed reactors. As well as specifically developed pyrolysis/gasification technologies, standard incineration technologies (i.e. grates, fluidised beds, rotary kilns, etc) may be adapted to be *operated* under pyrolytic or gasifying conditions i.e. with reduced oxygen levels *(sub-stoichiometric)*, or at lower temperatures. Often pyrolysis and gasification systems are coupled with downstream combustion of the *syngas* generated on combination processes. As well as the normal targets of waste incineration (i.e. effective treatment of the waste), the additional aims of gasification and pyrolysis processes are to:

- Convert certain fractions of the waste into process gas (called syngas);
- reduce gas cleaning requirements by reducing flue-gas volumes.

Both pyrolysis and gasification differ from incineration in that they may be used for recovering the chemical value from the waste (rather than its energetic value). The chemical products derived may in some cases then be used as feedstock for other processes. However, when applied to wastes, it is more common for the pyrolysis, gasification and a combustion based process to be combined, often on the same site as part of an integrated process. When this is the case the installation is, in total, generally recovering the energy value rather than the chemical value of the waste, as would a normal incinerator.

The following systems and concepts, listed in table 3, have been developed (with different levels of proven success on an industrial scale).

Table 3: Systems for pyrolysis and gasification

Pyrolysis – incineration systems for wastes	
System 1	Pyrolysis in a rotary kiln – coke and inorganic matter separation – incineration of pyrolysis gas
System 2	Pyrolysis in a rotary kiln – separation of inert materials – combustion of the solid carbon rich fraction and the pyrolysis gas
System 3	Pyrolysis in a rotary kiln – condensation of pyrolysis gas components - incineration of gas, oil and coke
System 4	Pyrolysis on a grate – directly connected incineration
System 5	Pyrolysis on a grate (with subsequent melting furnace for low metal content molten bottom ash production) – circulating fluidised bed (burnout of particles and gas).
Gasification system for wastes	
System 1	Fixed bed gasifier – pretreatment drying required for lumpy material
System 2	Slag bath gasifier – as fixed bed but with molten bottom ash discharge
System 3	Entrained flow gasifier – for liquid, pasty and fine granular material that may be injected to the reactor by nozzles
System 4	Fluidised bed gasifier – circulating fluid bed gasifier for pretreated municipal waste, dehydrated sewage sludge and some hazardous wastes
System 5	Bubbling bed gasifier – similar to bubbling fluidised bed combustors, but operated at a lower temperature and as a gasifier.

Source: IPPC 2006

Table 4: Different system for pyrolysis and gasification

Pyrolysis – gasification systems for wastes	
System 1	Conversion process – pyrolysis in a rotary kiln – withdrawal and treatment of solid phase – condensation of gas phase – subsequent entrained flow gasifier for pyrolysis gas, oil and coke
System 2	Combined gasification-pyrolysis and melting – partial pyrolysis in a push furnace with directly connected gasification in packed bed reactor with oxygen addition (e.g. Thermoselect).

Source: IPPC 2006

Other systems have been developed for the purpose of pretreating wastes that are then combusted in other industrial plants. Because of the high variety of applied techniques a more detailed description of the pyrolysis and gasification processes is abandoned.

2.3 Energy recovery

The characteristics of the waste delivered to the installation will determine the techniques that are appropriate and the degree to which energy can be effectively recovered. Both chemical and physical characteristics are considered when selecting processes. The chemical and physical characteristics of the waste actually arriving at plants or fed to the incinerator can be influenced by many local factors including:

- Contracts with waste suppliers (e.g. industrial waste added to MSW);
- on-site or off-site waste treatments or collection/separation regimes;
- market factors that divert certain streams to or from other forms of waste treatment.

In some cases the operator will have very limited scope to influence the characteristics of the waste supplied, in other cases this is considerable. The following table gives typical net calorific value ranges for selected waste types:

Table 5: Ranges and typical net calorific values for some incineration input wastes

Input type	Comments and examples	NCV in original substance (humidity included)	
		Range GJ/t	Average GJ/t
Mixed municipal solid waste (MSW)	Mixed household domestic wastes	6,3-10,5	9
Bulky waste	e.g. furniture etc delivered to MSWIs	10,5-16,8	13
Waste similar to MSW	Waste of a similar nature to household waste but arising from shops, offices etc.	7,6-12,6	11
Residual MSW after recycling operations	Screened out fractions from composting and materials recovery processes	6,3-11,5	10
Commercial waste	Separately collected fractions from shops and offices etc.	10-15	12.5
Packaging waste	Separately collected packaging	17-25	20
RDF-refuse derived fuels	Pellet or floc material produced from municipal and similar non-hazardous waste	11-26	18
Product specific industrial waste	e.g. plastic or paper industry residues	18-23	20
Hazardous waste	Also called chemical or special wastes	0,5-20	9,75
Sewage Sludges	Arising from waste water treatment works	See below	See below
	Raw (dewatered to 25% dry solids)	1,7-2,5	2,1
	Digested (dewatered to 25% dry solids)	0,5-1,2	0,8

Source: IPPC 2006

In addition to waste quality and technical aspects, the possible efficiency of a waste incineration process is influenced to a large extent by the output options for the energy produced. Processes with the option to supply electricity, steam or heat will be able to use more of the heat generated during the incineration for this purpose and will not be required to cool away the heat, which otherwise results in reductions in efficiency. The highest waste energy utilisation efficiency can usually be obtained where the heat recovered from the incineration process can be supplied continuously as district heat, process steam etc., or in combination with electricity generation. However, the adoption of such systems is very dependent on plant location, in particular the availability of a reliable user for the supplied energy.

The generation of electricity alone (i.e. no heat supply) is common, and generally provides a means of recovering energy from the waste that is less dependent on local circumstances. The following table 6 gives approximate ranges for the *potential* efficiencies at incineration plants in a variety of situations. The actual figures at an individual plant will be very site-specific. The idea of the table is therefore to provide mean values to compare what might be achievable in favorable circumstances. Doubts of calculation methods also make figures hard to compare – in this case the figures do not account for boiler efficiencies (typical losses ~20%), which explains why some figures approach 100% (figures exceeding 100% are also quoted in some cases).

Table 6: Energy conversion efficiencies of different types of WI plants

Plant type	Reported potential thermal efficiency% ((heat + electricity) /energy output from the boiler)
Electricity generation only	17-30
Combined heat and power plants (CHP)	70-85
Heating stations with sales of steam and/or hot water	80-90
Steam sales to large chemical plants	90-100
CHP and heating plants with condensation of humidity in flue-gas	85-95
CHP and heating plants with condensation and heat pumps	90-100

Note: The figures quoted in this table are derived from simple addition of the MWh of heat and MWh electricity produced, divided by the energy output from the boiler. No detailed account is taken of other important factors such as: process energy demand (support fuels, electrical inputs); relative CO_2 value of electricity and heat supply (i.e. generation displaced).

Source: IPPC 2006

The potential efficiencies are dependent on self-consumption of heat and electricity. Without taking the self-consumption into account, the calculated efficiencies of some facilities can lead to figures quoted of over 100%. Distortions of efficiency figures are also common when boiler heat exchange losses are discounted (i.e. a boiler efficiency of 80% means that 20% of the flue-gas heat is not transferred to the steam, sometimes efficiency is quoted in relation to the heat transferred to the steam rather than the heat in the waste). Where there is no external demand for the energy, a proportion is often used on-site to supply the incineration process itself and thus to reduce the quantity of imported energy to very low levels. For municipal plants, such internal use may be in the order of 10% of the energy of the waste incinerated.

Cooling systems are employed to condense boiler water for return to the boiler. Processes that are conveniently located for connection to energy distribution networks (or individual synergistic energy users) increase the possibility that the incineration plant will achieve higher overall efficiencies (IPPC 2006).

2.4 Flue-gas treatment (FGT) treatment systems

Flue-gas treatment systems are constructed from a combination of individual process units that together provide an overall treatment system for the flue-gases. A description of the individual process units, organised according to the substances upon which they have their primary effect, is given in this chapter. FGT systems are essential to remove substances that are harmful to humans and the environment (e. g. CO_2, NO_x, SO_x aromatic substances, etc.) from the flue gas. The FGT makes a big part in modern incineration plants and plays a crucial role in the discussion of climate change and green house gas emissions.

Figure 7: Overview of potential combinations of FGT systems

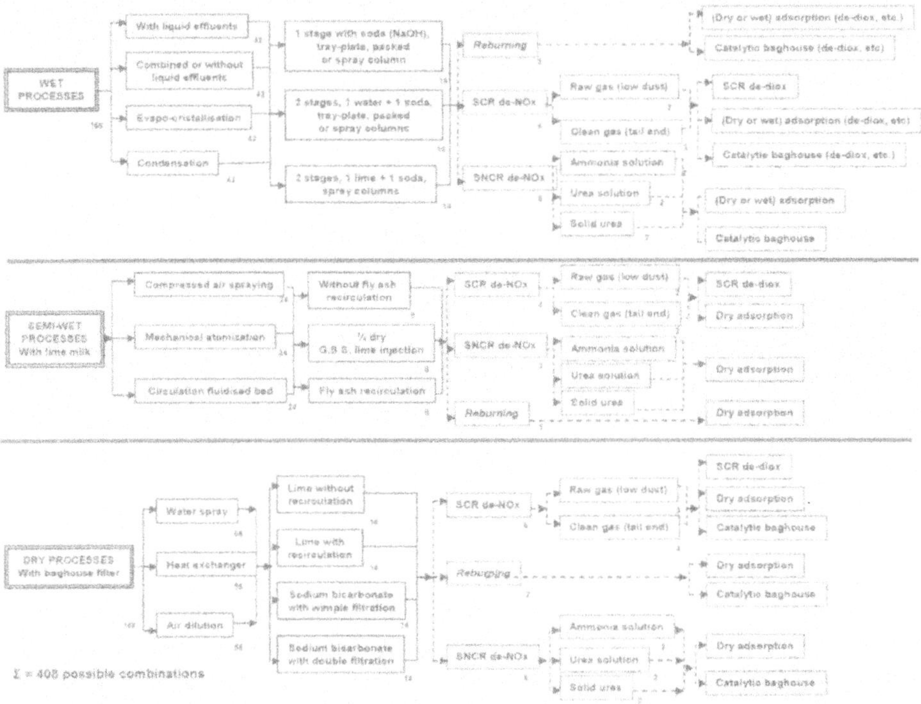

Source: IPPC 2006

2.4.1 Electrostatic Precipitators

Figure 8: Electrostatic precipitators (electrostatic filters)

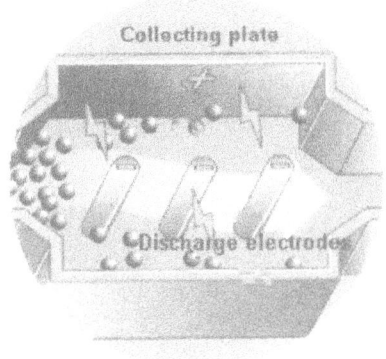

Source: IPPC 2006

Electrostatic precipitators are sometimes also called electrostatic filters. The efficiency of dust removal of electrostatic precipitators is mostly influenced by the electrical resistivity of the dust. If the dust layer resistivity rises to values above approx. 10^{11} to $10^{12}\,\Omega$ cm removal efficiencies are reduced. The dust layer resistivity is influenced by waste composition. It may thus change rapidly with a changing waste composition, particularly in hazardous waste incineration. Sulphur in the waste (and water content at operational temperatures below 200°C often reduces the dust layer resistivity as SO_2 (SO_3) in the flue-gas and therefore facilitates deposition in the electric field.

Wet electrostatic precipitators are based upon the same technological working principle as electrostatic precipitators. With this design, however, the precipitated dust on the collector plates is washed off using a liquid, usually water. This may be done continuously or periodically. This technique operates satisfactorily in cases where moist or cooler flue-gas enters the electrostatic precipitator.

2.4.2 Condensation electrostatic precipitators

The condensation electrostatic precipitator is used to deposit very fine, solid, liquid or sticky particles, for example, in the flue-gas from hazardous waste incineration plants. Unlike conventional wet electrostatic precipitators, the collecting surfaces of condensation electrostatic precipitators consist of vertical plastic tubes arranged in bundles, which are externally water-cooled. The dust-containing flue-gas is first cooled down to dew-point temperature in a quench by direct injection of water and then saturated with vapour. By cooling the gases in the collecting pipes further down, a thin, smooth liquid layer forms on the inner surface of the tubes as a result of condensation of the vapour. This is electrically earthed and thus serves as the passive electrode.

Figure 9: Condensation electrostatic precipitator

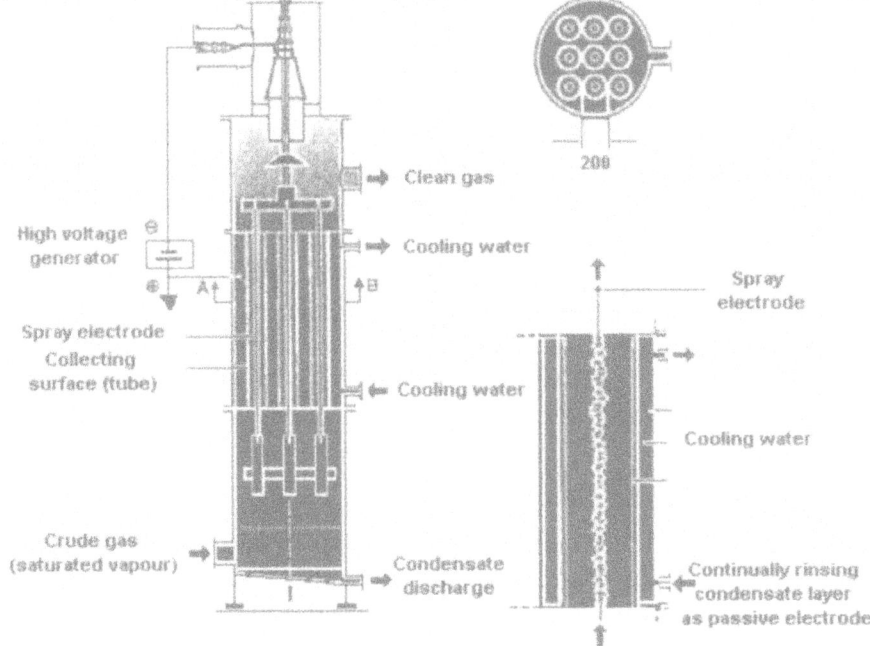

Source: IPPC 2006

2.4.3 Ionisation Wet Scrubbers (IWS)

The purpose of the Ionisation Wet Scrubber (IWS) is to remove various pollutants from the flue-gas flow. The IWS combines the principles of electrostatic charging of particles, electrostatic attraction and deposition for aerosols

- (smaller than 5 µm);
- vertical deposition for coarse, liquid and solid particles (larger than 5 µm);
- absorption of hazardous, corrosive and malodorous gases.

The IWS system is a combination of an electrostatic filter and a packed scrubber. It is reported to require little energy and has a high deposition efficiency for particles in the submicron as well as the micron range. A high voltage zone is installed before each packed tower stage. The function of the high voltage zone is to ionise the particles (dust, aerosols, submicron particles) contained in the fluegas. The negatively charged particles induce opposing charges on the neutral surface of the wetted packing material and the falling water drops. Because of this they are attracted and are then washed out in the packed section. This is referred to as Image/Force attraction (IF attraction), i.e. attraction through electron shift. Hazardous, corrosive and malodorous gases are also absorbed in the same scrubber fluid and chemically combined to be discharged with the scrubber effluent. Another type of ionization wet scrubber includes a Venturi. The pressure changes that occur through the Venturi allows the fine particles to grow and the electrode charges them. They are then collected by the dense layer of water droplets projected by a nozzle, serving as collecting electrode

2.4.4 Fabric Filters

Material which has to be separated is often fed on a belt conveyor. The conveyor usually operates at fast velocities so that its function is almost like an isolating device.

Figure 10: Fabric / Bag Filter

Source: IPPC 2006

Fabric filters, also called bag filters, are very widely used in waste incineration plants. Filtration efficiencies are very high across a wide range of particle sizes. At particle sizes below 0,1 microns, efficiencies are reduced, but the fraction of these that exist in the flue-gas flow from waste incineration plants is relatively low. Low dust emissions are achieved with this technology. It can also be used following an ESP and wet scrubbers.

Compatibility of the filter medium with the characteristics of the flue-gas and the dust, and the process temperature of the filter are important for effective performance. The filter medium should have suitable properties for thermal, physical and chemical resistance (e.g. hydrolysis, acid, alkali, oxidation). The gas flowrate determines the Mechanical and thermal stress on the filter material determines service life, energy and maintenance requirements. appropriate filtering surface i.e. iltering velocity.

2.4.5 Cyclones and Multi-Cyclones

Cyclones and multi-cyclones use centrifugal forces to separate particulate matter from the gas stream. Multi-cyclones differ from single cyclones in that they consist of many small cyclone units. The gas flow enters the separator tangentially and leaves from a central port. Solids are forced to the outside of the cyclone and collected at the sides for removal.

In general, cyclones on their own cannot achieve the emission levels now applied to modern waste incinerators. They can, however, have an important role to play where applied as a prededuster before other flue-gas treatment stages. Energy requirements are generally low as there is no pressure drop across the cyclone. Advantages of cyclones are their wide operational temperature range and robust construction. Erosion of cyclones, particularly at the point of impingement of dirty flue-gases, can be an issue where the flue-gas is more heavily loaded with particulate, and particularly where bed material escapes from fluidised bed plants. Circulating fluidised beds usually incorporate a cyclone for the removal and recirculation of the bed material to the furnace (IPPC 2006).

2.4.6 Techniques to further clean the flue gas

In the following will be described which techniques are applied to reduce concentration of acids, organic carbons, nitrogen-oxides and greenhouse gases (e.g. CO_2) in the flue gas. This chapter gives a brief overview of technical solutions that are applied successfully.

Sulphur dioxide and gaseous halogens are cleaned from flue-gases by the injection of chemical or physical sorption agents, which are brought into contact with the flue gas. According to technique, the reaction products are dissolved or dry salts.

2.4.6.1 Reduction of sulphur dioxides and halogens

To reduce content of sulphur dioxides and halogens in the flue gas are generally three techniques present that can be divided in dry, semi-wet/dry and wet sorption systems.

Table 7:   Further flue gas cleaning techniques

Dry systems	In dry sorption processes the absorption agent (usually lime or sodium bicarbonate) is fed into the reactor as a dry powder. The dose rate of reagent may depend on the temperature as well as on reagent type. With lime this ratio is typically two or three times the stoichiometric amount of the substance to be deposited, with sodium bicarbonate the ratio is lower. This is required to ensure emission limits are complied with over a range of inlet concentrations. The reaction products generated are solid and need to be deposited from the flue-gas as dust in a subsequent stage, normally a bag filter.
Semi-dry / semi-wet systems	These are also called *semi-dry* processes. In the spray absorption, the absorption agent is injected either as suspension or solution into the hot flue-gas flow in a spray reactor. This type of process utilises the heat of the flue-gas for the evaporation of the solvent (water). The reaction products generated are solid and need to be deposited from the flue-gas as dust in a subsequent stage e.g. bag filter. These processes typically require overdoses of the sorption agent of 1,5 to 2,5.
Wet systems	Wet systems use different types of scrubbers like jet, rotation, dry tower, venture and packed tower scrubbers. The scrubber solution is (in the case of water only injection) strongly acidic (typically pH 0-1) due to acids forming in the process of deposition. HCl and HF are mainly removed in the first stage of the wet scrubber. The effluent from the first stage is recycled many times, with small fresh water addition and a bleed from the scrubber to maintain acid gas removal efficiency. In this acidic medium, deposition of SO_2 is low, so a second stage scrubber is required for its removal. Removal of sulphur dioxide is achieved in a washing stage controlled at a pH close to neutral or alkaline (generally pH 6-7) in which caustic soda solution or lime milk is added. For technical reasons this removal takes place in a separated washing stage, in which, additionally occurs further removal of HCl and HF.

Source: IPPC 2006

A special case of reduction of sulphur is the *direct desulphurization*. Desulphurisation in fluidised bed processes can be carried out by adding absorbents (e.g. calcium or calcium/magnesium compounds) directly into the incineration chamber. Additives such as limestone dust, calcium hydrate and dolomitic dust are used. The system can be used in combination with downstream flue-gas desulphurisation.

The arrangement of the jets and the injection speed influence the distribution of the absorbents and thus the degree of sulphur dioxide deposition. Part of the resulting reaction products are removed in filter installations downstream; however, a significant proportion remains with the bottom ashes. Therefore, direct desulphurisation may impact on bottom ash quality. Ideal conditions for direct desulphurisation exist in a cycloid furnace due to the constant temperature level (IPPC 2006).

2.4.6.2 Reductions of emissions of nitrogen oxides (NO_x)

Basically NO_x can be build in three different ways. During combustion a part of the air nitrogen is oxidised to nitrogen oxides, so called *thermal NO_x*. Secondly nitrogen in the fuel is oxidized to nitrogen oxides, so called *fuel NO_x* and thirdly atmospheric nitrogen can also be oxidised by reaction with CH radicals and intermediate formation of HC (formation of NO_x via radical reaction, so called *prompt NO_x*. (IPPC 2006). The amount of built nitrogen depends on the different temperature levels during combustion and in the air. Nitrogen oxides have evidently influence on the ozone concentration in the atmosphere which again can lead to diseases. Another effect of NO_x is the so called acid rain which damages flora and fauna.

Figure 11: Temperature Dependence on various NO_x formation mechanisms during WI

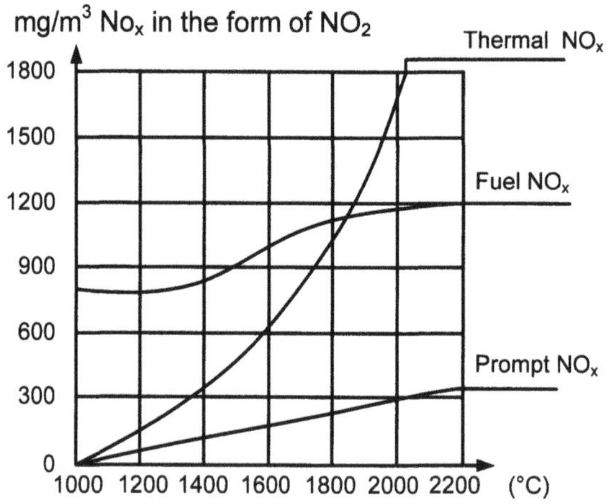

Source: IPPC 2006

Generally two different kinds of techniques are distinguished, primary and secondary techniques. The following table gives a brief overview.

Table 8: Primary techniques for reduction of nitrogen oxides (NO_x)

Air supply, gas mixing and temperature control	The use of a well distributed primary and secondary air supply to avoid the uneven temperature gradients that result in high temperature zones and, hence, increased NO_x production is a widely adopted and important primary measure for the reduction of NO_x production. Although sufficient oxygen is required to ensure that organic materials are oxidised (giving low CO and VOC emissions), the overfeeding of air can result in additional oxidation of atmospheric nitrogen, and the production of additional NO_x.

Flue-Gas Recirculation (FGR)	This technique involves replacement of around 10-20% of the secondary combustion air with recirculated flue-gases. NOx reduction is achieved because the supplied re-circulated flue-gases have lower oxygen concentration and therefore lower flue-gas temperature which leads to a decrease of the nitrogen oxide levels.
Oxygen injection	The injection of either pure oxygen or oxygen enriched air provides a means to supply the oxygen required for combustion, while reducing the supply of additional nitrogen that may contribute to additional NOx production.
Staged combustion	Staged combustion has been used in some cases. This involves reducing the oxygen supply in the primary reaction zones and then increasing the air (and hence oxygen) supply at later combustion zones to oxidise the gases formed. Such techniques require effective air/gas mixing in the secondary zone to ensure CO (and other products of incomplete combustion) are maintained at low levels.
Natural gas injection (re-burn)	Natural gas injection into the over-grate region of the furnace can be used to control NO_X emissions from the combustor. For MSWIs, two different natural gas based processes have been developed: • Re-burning – a three stage process designed to convert NO_X to N_2 by injecting natural gas into a distinct re-burn zone located above the primary combustion zone • Methane de-NO_x – this technique injects natural gas directly into the primary combustion unit to inhibit NOx formation.
Injection of water into furnace/flame	A properly designed and operated injection of water either into the furnace or directly into the flame can be used to decrease the hot spot temperatures in the primary combustion zone. This drop in peak temperature can reduce the formation of thermal NO_x.

Source: IPPC 2006

The Directive 2000/76/EC requires a daily average NO_x (as NO_2) clean gas value of 200 mg/Nm³. In order to achieve compliance at this level, it is common for secondary measures to be applied. For most processes the application of ammonia or derivatives of ammonia (e.g. urea) as reduction agent has proved successful. The nitrogen oxides in the flue-gas basically consist of NO and NO_2 and are reduced to nitrogen N_2 and water vapor by the reduction agent. Two processes are important for the removal of nitrogen from flue-gases – the Selective Non-Catalytic Reduction (SNCR) and the Selective Catalytic Reduction (SCR).

Figure 12: Secondary techniques for reduction of nitrogen oxides (NO$_x$)

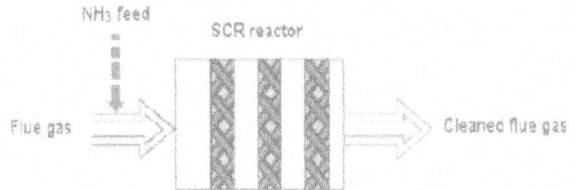

Source: IPPC 2006

Selective Catalytic Reduction (SCR) is a catalytic process during which ammonia mixed with air (the reduction agent) is added to the flue-gas and passed over a catalyst, usually a mesh (e.g. platinum, rhodium, TiO$_2$, zeolites). When passing through the catalyst, ammonia reacts with NO$_x$ to give nitrogen and water vapor.

Figure 13: Secondary techniques for reduction of nitrogen oxides (NO$_x$)

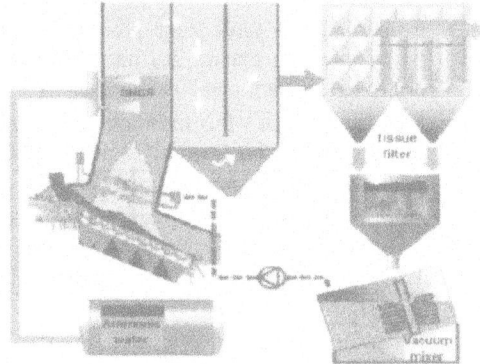

Source: IPPC 2006

In the Selective Non-Catalytic Reduction (SNCR) process nitrogen oxides (NO + NO$_2$) are removed by selective non-catalytic reduction. With this type of process the reducing agent (typically ammonia or urea) is injected into the furnace and reacts with the nitrogen oxides. The reactions occur at temperatures between 850 and 1000°C, with zones of higher and lower reaction rate within this range.

Reducing NO$_x$ by more than 60-80% via SNCR requires a higher addition of the reducing agent. This can lead to emissions of ammonia, also known as *ammonia slip*.

2.4.6.3 Reduction of heavy metals

Heavy metals (except mercury) in incineration are converted mainly into non-volatile oxides and deposited with flue ash. Thus, the main techniques of relevance are, therefore, those applicable to dust removal (see 2.4). Activated carbon is reported to be also used for reducing heavy metals emissions. Mercury needs special treatment methods which are not mentioned in this paper.

2.4.6.4 Reduction of organic carbon compounds

The following table gives an overview of applied techniques for the reduction of organic carbon compounds (i.e. halogenated aromatic hydrocarbons, polycyclic aromatic hydrocarbons (PAH), benzene, toluene and xylene (BTX), PCDD/F).

Table 9: Applied techniques for reduction of organic carbon compounds

Adsorption on activated carbon reagents in an entrained flow system	Activated carbon is injected into the gas flow. The carbon is filtered from the gas flow using bag filters. The activated carbon shows a high absorption efficiency for mercury as well as for PCDD/F. Different types of activated carbon have different adsorption efficiencies. This is believed to be related to the specific nature of the carbon particles, which are, in turn, influenced by the manufacturing process.
SCR systems	SCR systems are used for NO_x reduction. They also destroy gaseous PCDD/F (not particle bound) through catalytic oxidation; however, in this case, the SCR system must be designed accordingly, since it usually requires a bigger, multi-layer, SCR system than for just the de-NO_x function. Destruction efficiencies for PCDD/F of 98 to 99.9% are seen.
Catalytic bag filters	Filter bags are either impregnated with a catalyst, or the catalyst is directly mixed with organic material in production of fibres. Such filters have been used to reduce PCDD/F emissions.
Re-burn of carbon adsorbents	Carbon is used to adsorb dioxins (and mercury) at many waste incinerators. Where processes have another outlet for the mercury that provides an adequate removal rate, (i.e. a greater rate than the input rate to avoid circulation and hence emission breakthrough) it is possible for the net dioxin emissions from the plant to be reduced by re-burning the adsorbed PCDD/F by re-injection into the furnace.
Use of carbon impregnated plastics for PCDD/F adsorption	Plastics are widely used in the construction of flue-gas cleaning equipment due to their excellent corrosion resistance. PCDD/F is adsorbed on these plastics in wet scrubbers, where the typical operational temperature is 60–70 °C. If the temperature is increased by only a few degrees Celsius, or if the dioxin concentration in the gas is reduced the absorbed PCDD/F can be desorbed to the gas phase and increase emissions to air.
Static bed filters	Activated coke moving bed filters are used as a secondary cleaning process in the flue-gas of municipal and hazardous waste incineration. Using this adsorption system, it is possible to deposit substances contained in the flue-gas at extremely low concentrations with high efficiency. Lignite coke produced in hearth furnace coke process is used in moving bed absorbers.
Rapid quenching of flue-gases	This technique involves the use of a water scrubber to cool flue-gases directly from their combustion temperature to below 100 °C. The technique is used in some HWI. The action of rapid quenching reduces the residence of flue-gases in temperature zones that may give rise to additional de-novo PCDD/F synthesis.

Source: IPPC 2006

2.4.6.5 Reduction of greenhouse gases (CO_2, N_2O)

To decrease emissions of CO_2 to air exist three possible options to achieve this goal. Firstly optimizing energy recovery and supply of the incineration plant and controlling CO_2 emissions via an effective FGT. The Production of sodium carbonate by reacting CO_2 in the flue-gases with NaOH is possible but not further described in this paper.

To decrease emissions of N_2O, the following techniques are being used:

- Reduction of SNCR reagent dosing by SNCR process optimisation;
- selecting optimised temperature window for SNCR reagent injection;
- use of flow modeling methods to optimise injection nozzle locations;
- designing to ensure effective gas/reagent mixing in the appropriate temperature zone;
- over-stoichiometric burnout zones to ensure oxidation of nitrous oxide;
- utilization of ammonia instead of urea in SNCR.

The optimum temperature for the simultaneous minimisation of both NO_x and N_2O production is reported to be in the range 850–900°C. Under conditions where the temperature in the post combustion chamber is above 900 °C the N_2O emissions are reported to be low. N_2O emissions from the use of SCR are also low. Thus, provided combustion temperatures are above 850°C, in general, SNCR represents the only significant source of N_2O emissions at modern waste incinerators (IPPC 2006).

3. Emissions and consumptions

The main emissions of a WI plant are, due to the combustion process, air emissions, described in table 10:

Table 10: Emissions to air by an waste incineration plant

Carbon monoxide	**Carbon dioxide**	**Methane**
Total organic carbon	Polychlorinated dibenzo-dioxins and furans (PCDD/F)	Ammonia
Hydrogen chloride	Polychlorinated biphenyls (PCB)	Polyaromatic hydrocarbons
Hydrogen fluoride	Sulphur oxides	Nitrogen oxides
Hydrogen iodide and iodine, hydrogen bromide and bromine	Dust	Mercury and mercury compunds
Cadmium and thallium compunds	Other heavy metal compounds	

Source: IPPC 2006

Emissions to water and emissions of solid residues are not enlarged on in this chapter. For further information concerning these topics please read the full BREF document edited by IPPC in 2006.

Energy consumption of a WI plant is strongly connected to the efficiency of the thermal stage and the energy recovery system. Usually a WI plant has on its input side waste, support fuels and electricity. Electricity and heat are on the output side of a WI plant. The higher the efficiency of energy recovery the lower the energy consumption of the plant. Aspects of energy recovery were mentioned in sector 2.3.

4. Techniques to consider in the determination of Best Available Techniques

Additional techniques are considered generally to have potential for achieving a high level of environmental protection in the industries. For example management systems, process-integrated techniques and end-of-pipe measures can have significant influence on plants efficiency and therefore on the environment . Prevention, control, design, management and re-cycling procedures as well as the re-use of materials and energy techniques must be considered to achieve the objectives of IPPC. Because it is not possible to be exhaustive and because of the dynamic nature of industry, it is possible that there may be additional techniques not mentioned but which may also be considered as Best Available Techniques (BAT).

5. Best Available Techniques

Best Available Techniques in the waste incineration sector depends on the targeted product and the required product quality, the type and appearance of the input material and economic factors like market situation for specific materials (e.g. prices for metals, glass, paper and wood).

Depending on the kind of waste incineration plant BAT is to combine aforementioned techniques (see section 2. Applied Techniques) in such way that the energy recovery rate reaches the maximum and that emissions to air, water and soil are minimal. It is not possible to define BAT for one special kind of incineration plant, because all the techniques are adequate for different wastes/fuels so that only the well thought combination of these techniques can lead to an environmentally and economic benefit.

Management systems like environmental management systems, raw material and utility management systems, knowledge of waste IN and waste OUT and therefore management of waste analysis are also considered to be BAT for a waste incineration plant. For further information concerning BAT in waste treatment methods please consult the BAT for waste treatment installations.

6. Emerging Techniques

Emerging technique is understood in this document as a novel technique that has not yet been applied in any industrial sector on a commercial basis. This chapter contains those techniques that may appear in the near future and that may be applicable to the waste incineration sector. Emerging techniques in the waste incineration sector concentrates on optimisations in the combustion and FGT process. In this paper a short overview will be given what is considered for future technologies. Because of the high amount of emerging techniques and their state of development these techniques will just be listed and described very briefly.

The following table shows the techniques which are in development status and considered to bring economic and ecologic improvements in waste incineration sector.

Table 11: Emerging techniques in WI sector

Technique	Brief description
Use of steam as a spraying agent in post combustion chamber burners instead of air	Technical description see EURITS (2003). "List of techniques for consideration as BAT", EURITS.
Application involving the reheating of turbine steam	Another option to increase the efficiency of electricity production is the reheating of turbine steam after its first passage through the turbine. For this application, steam temperature is limited to 400°C, but steam pressure increases. The figure below provides a simplified process scheme for this option. Efficiency increase of about 2-3% is achievable with this technique.
Measures in the crude flue-gas area for reducing dioxin emissions	A reduction in dioxins can be achieved through the following measures in the crude flue-gas area, which seek to reduce dioxin formation by inhibiting the reactions or reducing the presence of dusts in the temperature range 450-200°C. 1st The addition of inhibitors to the waste – efficiency is limited and secondary reactions require. Consideration. 2nd employment of hot gas dedusters – so far only little experience from pilot tests: dedusting using ceramic filters or cyclones at temperatures of approx. 800°C; dedusting at temperatures above 450°C e.g. with hot gas electrostatic filters. Third reduction of deposits of airborne dust on the flue-gas path by effective cleaning of flue-gas vents, boiler, heating plates – a well proven maintenance related issue.

Technique	Brief description
Oil scrubber for the reduction of poly-halogenated aromatics and polyaromatic hydrocarbons (PAHs) in the flue-gases from incineration plants	Dioxins and furans have very low solubility in water (they are more lipophilic) and therefore they are not removed in wet scrubbers to a significant and reliable extent. Any removal which does take place is generally due to the removal of PCDD/F that is adsorbed onto particulate matter removed in the wet scrubber. A high boiling partly unsaturated oil or a oil-water emulsion of such oil therefore provide suitable scrubbing media.
Use of CO_2 in flue-gases for the production of sodium carbonate	If the flue-gas is brought into contact with caustic soda solution, the carbon dioxide reacts with the sodium hydroxide to form sodium carbonate. The liquid is odorless and colorless. The carbonate solution may be used as a raw material, e.g. chemical plants, paper industry.
Increased bed temperature, combustion control and oxygen addition in a grate incinerator	The basic concept of this process (known as the SYNCOM plus process) is the integrated sintering of ash in the waste bed of a grate based energy from the waste incinerator.
$FeSO_4$ stabilisation of FGT residues	This stabilisation involves a five-step procedure, where the residues are first mixed with a $FeSO_4$ solution and then aerated with atmospheric air at liquid/solid ratio of 3 l/kg, in order to oxidise Fe(II) to Fe(III) and precipitate iron oxides. This step also includes extraction of soluble salts.
CO_2 stabilisation of FGT residues	This stabilisation process involves a two-step procedure. The residues are first washed in order to extract soluble salts, and then dewatered and washed again in a plate and frame filter press The residues are then re-suspended, and CO_2 and/or H_3PO_4 is added. Finally, the residues are dewatered again and washed at the filter press.
Combined dry sodium bicarbonate + SCR + scrubber FGT systems	This technique consists in combining dry FGT with sodium bicarbonate with a SCR system and a scrubber. As sodium bicarbonate presents a wide operating temperature range (140-300°C) and leads to SO_X emissions below 20 mg/Nm₃ (SO_3 included), it ideally combines with an SCR without reheating the FG; although FG reheat maybe needed if stack temperature is too low after wet scrubber.

Source: IPPC 2006

7. References

(IPPC 2006)   European Commission, Integrated Pollution Prevention and Control, "Reference Document on Best Available Techniques for the Waste Treatment Industries", August 2006

(TCA 2007)   Turkish Court of Accounts, "Waste Management in Turkey, National regulations and implementation results", Performance Audit Report, January 2007

Chapter 7

Best Available Techniques – Working Paper – Waste Management

Kerstin Kuchta, Konstantin Haker

The structure of this document is also derived from an original reference document on Best Available Techniques (BAT). The determination of BAT is used as a tool to display the state-of-the-art of waste management technologies with the generally accepted guidelines of the IPCC Directive 2008(1/EC).

Scope

The scope of this document is the description of different treatment methods for Refuse Derived Fuel (RDF) production from wastes that are successfully used in practice and which have proved value in industry. In general this paper distinguishes between mechanical and biological treatment methods. Matters of landfill, incineration, the treatment of liquid wastes like solvents and oil, waste water treatment as well as sewage sludge incineration are not considered in this paper.

1. General Information

1.1 The purpose of waste management

Secondary products are inherent to any industrial process and normally cannot be avoided. In addition, the use of products by society leads to residues. In many cases, these types of materials (both secondary products and residues) cannot be re-used by other means and may become not marketable. These materials are typically given to third parties for further treatment.

The reason for treating waste is not always the same and often depends on the type of waste and the nature of its subsequent fate. Some waste treatments and installations are multipurpose. In this document, the basic reasons for treating waste are:

- To reduce the hazardous nature of the waste;
- to separate the waste into its individual components, some or all of which can then be put to further use/treatment;

- to reduce the amount of waste which has to be finally sent for disposal;
- to transform the waste into an useful material.

The waste treatment processes may involve the displacement and transfer of substances between media. For example, some treatment processes result in a liquid effluent sent to sewer and a solid waste sent to landfill, and others result in emissions to air mainly due to incineration. Alternatively, the waste may be rendered suitable for another treatment route, such as in the combustion of recovered fuel oil. There are also a number of important ancillary activities associated with treatment, such as waste acceptance and storage, either pending treatment on site or removal off site.

1.2 Waste Management

In the sector of waste management exist a high variety of different types of treatment plants that target a wide range of process products and therefore treat a wide variety of wastes. Because of the increasing shortage of raw materials like for instance metals and wood (paper industry), the management of wastes and as a result of the production of secondary raw materials are getting more and more important in two ways, environmentally and economically.

The processes and activities found in the WT sector are divided into four sections in this document. Such structure/classification should not be interpreted as any attempt to interpret IPPC Directive or any EC waste legislation. These are:

- *Common techniques*. This covers those stages found in the waste sector that are generally applied and that are not specific to any individual type of waste treatment (e.g. reception, blending, sorting, storage, energy system, management). The unit operations associated with these treatments are also covered;
- *biological treatments* and some mechanical-biological treatments (e.g. aerobic/anaerobic digestions). The unit operations associated to these treatments are also covered;
- treatments applied to turn a waste into a material that can be *used as a fuel* in different industrial sectors (e.g. refuse derived fuel (RDF) production);
- end-of-pipe techniques used in waste treatment installations for the *abatement of emissions*.

This paper contains a description of special treatment methods that use waste to produce a material that can be used as a fuel in different industrial sectors (RDF production). To understand which kind of wastes show a applicability to produce RDF it is necessary to know which types of incineration plants exist. The incineration sector can approximately be divided in three different parts. These are:

a) *Mixed municipal waste incineration* – treating typically mixed and largely untreated household and domestic wastes but may sometimes include certain industrial and commercial wastes (industrial and commercial wastes are also separately incinerated in dedicated industrial or commercial non-hazardous waste incinerators). RDF can be co-incinerated in such plants;

b) *pretreated municipal or other pretreated waste incineration* – installations that treat wastes that have been selectively collected, pretreated, or prepared in some way, such that the characteristics of the waste differ from mixed waste. Specifically prepared refuse derived fuel incinerators like co-combustion power plants and RDF power plants fall in this sub-sector;

c) *hazardous waste incineration* – this includes incineration on industrial sites and incineration at merchant plants (that usually receive a very wide variety of wastes that cannot be further treated). The production of RDF from hazardous wastes is not practised.

Basically, waste incineration is the oxidation of the combustible materials contained in the waste. Waste is generally a highly heterogeneous material, consisting essentially of organic substances, minerals, metals and water. During incineration, flue-gases are created that will contain the majority of the available fuel energy as heat. The organic substances in the waste will burn when they have reached the necessary ignition temperature and come into contact with oxygen. The actual combustion process takes place in the gas phase in fractions of seconds and simultaneously releases energy. Where the calorific value of the waste and oxygen supply is sufficient, this can lead to a thermal chain reaction and self-supporting combustion, i.e. there is no need for the addition of other fuels.

The production of RDF for incineration plants highly depends on for what kind of incineration the RDF is meant so that the end of pipe production of RDF needs different techniques depending on the type of waste treatment plant.

1.3 Waste Management in Turkey

Accordingly, the amount of wastes produced per person reaches to 2 kg daily and each person produces waste equal to 10 folds of his/her weight. 34 million tons of municipal waste and 17,5 million tons of industrial waste are produced in Turkey According to 2004 statistical figures of Turkish Statistical Institute (TURKSTAT).

Best practices in developed countries show that only 35-45% diversion of all waste is disposed into landfills; the rest of the waste is recycled and transferred to an economic asset. More than half of the waste generated could be reused and recycled, and transformed from a problem into an asset. However, it is a well-known fact that the rates of recycling are very low, although there are no hard data in this area. Since a sound waste management infrastructure is not established in Turkey, each and every year, millions of tons of natural resources, the employment opportunity for thousands of people, a wealth of millions of dollars are wasted and the revival capacity of environment is rapidly exhausted (TCA 2007).

2. Applied Techniques

Only an overview of technical equipment that is implemented in waste treatment plants can be given in this paper. Every treatment plant has its own and special waste processing system so that a resilient conclusion concerning BAT in waste progression procedures is impossible. Therefore applied techniques are listed and briefly described in the following.

2.1 Common techniques applied in the sector of WT

For most WT plants, the following structure for waste processing is relevant: *a) reception b) storage c) treatment d) storage of residues and dispatch* (see Figure 1). Each of the steps require knowledge and control of the waste as well as specific reception and processing management. Knowledge of wastes, before they are receipted and treated, is a key factor for the management of a WT plant. The aim of this section is to present the different types of controls and analyses which can be carried out during the waste treatment process, from the pre-receiption and arrival of the waste at the site, to the final dispatch for either recovery (energetic/substantial) or disposal on landfill sites of the waste. Figure 1 shows a flow diagram for a typical waste treatment installation.

Figure 1: Basic steps of waste management/treatment

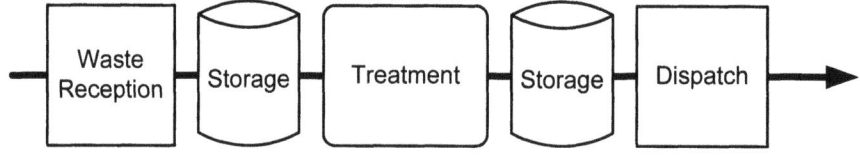

Source: IPPC 2006

In the following different treatment aggregates such as separation, crushing and screening units which are used at present especially in the german waste management/treatment sector are described. Waste treatment plants can be classified in three different groups of installations:

a) *Waste installations included in the same place where the waste is produced.* These typically serve a rather small number of wastes types and can provide only a restricted number of treatments;

b) *specific dedicated waste installations*, which may provide one or several operations but which typically treat only a small number of waste types or which produce a relatively small amount of output;

c) *integrated waste treatment installations*. Some waste treatment installations are not standalone installations only containing a single type of treatment. Some of them are designed to provide a wide variety of services, and they are designed to treat a great variety of waste types. Waste treatment installations are designed to produce required waste treatment services. For example, sometimes they are designed to provide a certain type of treatment to deal with a large amount and variety of different waste types (e.g. municipal solid wastes, aqueous wastes). Figure 2 is one example of such a complex installation.

Figure 2: Example of an integrated waste treatment installation

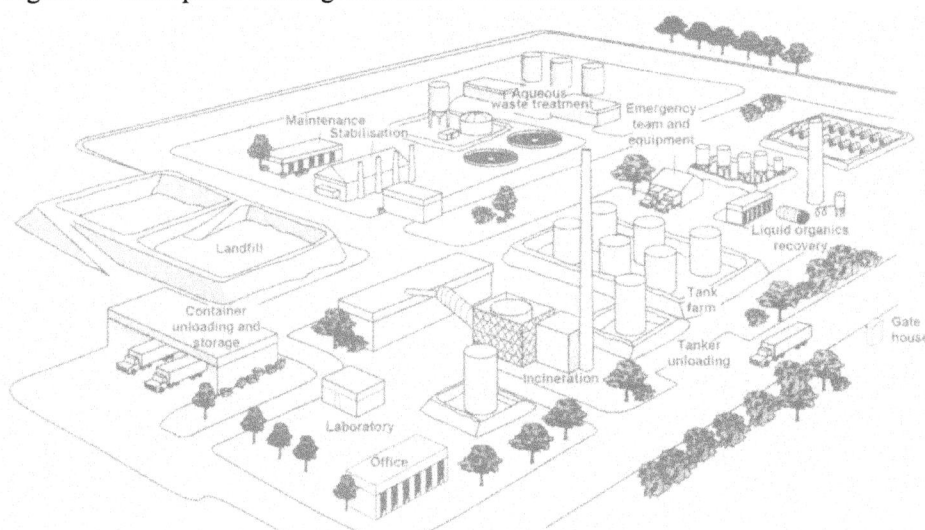

Source: IPPC 2006

2.2 Selection of techniques used for preparation of solid waste fuel

Following techniques are used for the preparation of solid waste fuel from non-hazardous wastes. Usually the incoming wastes are pre-sorted in different classes depending on the kind of waste and the treatment plant. In general following steps are commonly applied:

a) classifying solid waste by pre-sorting and afterwards crushing and screening the bulky fractions before the sorting operation;

b) applying magnetic separators and other separators e.g. non-ferrous and all metals separators, positive and negative sorting systems, automatic picking applications and near infrared spectroscopy;
c) carrying out the mixing and sieving operations in closed areas;
d) using nitrogen mixing devices to make the atmosphere inert when there is risk of explosion.

Hazardous wastes must be treated in a different way and are usually given to recovery or special hazardous waste incineration plants depending on the type of waste. Technique d. for example is applied for hazardous waste treatment with a risk of explosion.

2.3 Pre-sorting

Before the waste is proceeded to further treatment methods like the crushing unit it is roughly presorted. Pre-sorting can be realised by:

- A separate collection of wastes (e.g. MSW, plastics, electronical devices, metal containing wastes, paper, glas, wood, etc.);
- a separation of wastes on-site via excavators and other technical equipment or by hand. This is mainly used to extract hazardous substances (if present) or to separate material that could damage the technical equipment of the plant.

2.4 Crushing units

The following table shows examples for different crushing units used by waste treatment industry.

Table 1: Different crushing techniques

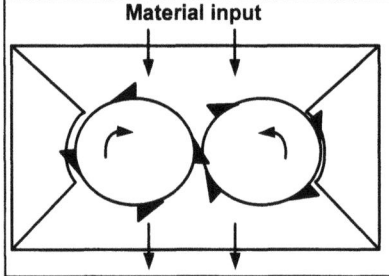	*Cutting mills* or granulators are used to crush soft to medium-hard materials and are mainly classified in mills that use blades and such that use cutters. It is used to pre-crush the waste to prevent the following treatment unit from damage and to create a fraction where contraries and materials with values can be separated more easily.

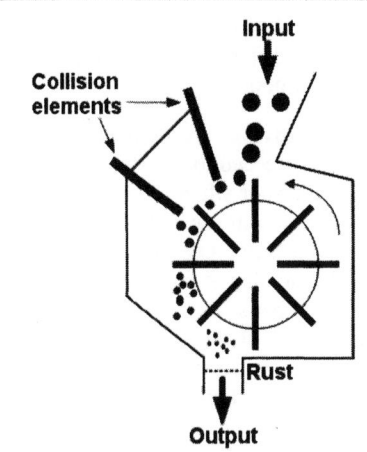

Primary Impact mills are mainly used for processing of medium-hard and hard materials.

After entering the crushing radius of the rotor, the feed material is grasped by the firmly fixed beater ledges and crushed by the force of its impact against the upper and lower collision elements. As soon as the material is crushed fine enough it passes the rust which has depending on connected treatment method different apertures. Crushing happens via smashing the material against the rotor cabin and/or the collision elements and/or against itself.

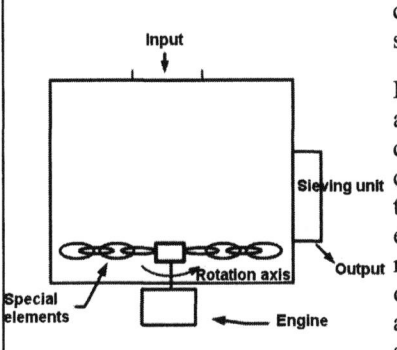

The *cross-flow-shredder* is used for hard materials (e.g. catalysts, metal scrap, electronic devices, cooling systems, freezers and RDF/MBA scrap.

It completely avoids the use of blades so that straight cuts are much better than with normal systems. It concentrates on decomposing the waste material, by colliding the accelerated feed material against itself, so that separating out the individual components and elements is avoided. In consequence broken down recyclable material is easily accessible while a leakage of any contaminants is avoided. The special elements are e.g. chains that are rotated and hit the material and smash against the cabin walls or against other pierces of the materials.

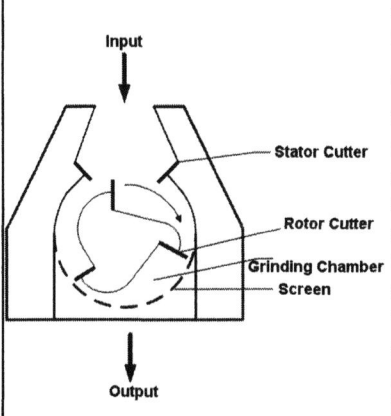

Size reduction in the cutting mill takes place by cutting and shearing forces.

The input material passes through the hopper and into the grinding chamber where it comes into contact with a rotor equipped with three cutting blades; it is comminuted between the rotor cutters and the stationary cutters inserted in the housing. The chamber dwell time is short; as soon as the material can pass through the openings of the bottom sieve it is discharged.

The cutting mill is suitable for size reduction of soft, medium-hard, elastic or fibrous materials, whose size can be reduced without requiring the use of extremely high forces. Furthermore exist bell mills, disc mills and mortar grinders that are not explained in this paper.

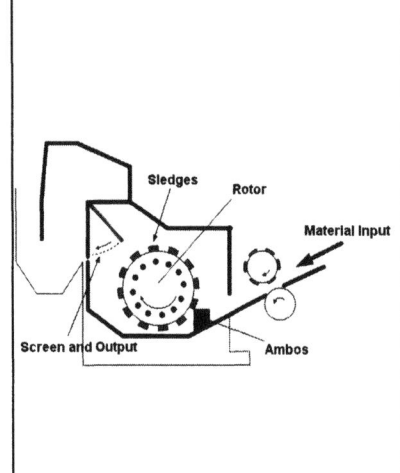

Crushing unit for solubilising the input material such as electronical devices, different kinds of scrap (cars, mixed scrap, etc.) and wood.

Two rolls on the belt conveyor precompress incoming material before it is proceeded to the shredder. Inside the shredder cabin is a rotor where sledges are fixed. Material is crushed while it is located between ambos and sledges. Inside the rotor cage material is flung with a high velocity so that it smashes against the cabin walls and other parts of the material so that crushing happens through sledges and impact forces. An optional screen with a defined diameter, which is usually installed on the bottom of the cabin (some have it on top like in the picture on the left), gives the grade of crushing. The sieving grad depends on the connected separation or treatment method.

Source: VDI 2009

The mentioned examples represent main types of treatment aggregates which are used for crushing in the waste management sector. Depending on the kind and composition of the wastes there may other techniques be more adequate than one of these shown above. For example exist a high variety of mills on the market, so that is has to be well balanced which type of mill is the optimum for the planned application.

2.5 Magnetic separation of ferrous metals

After the crushing unit there typically is a magnetic separator installed. The following Figure 3 gives an overview of the different types and installation forms of a magnetic separator unit.

Figure 3: Magnetic separation

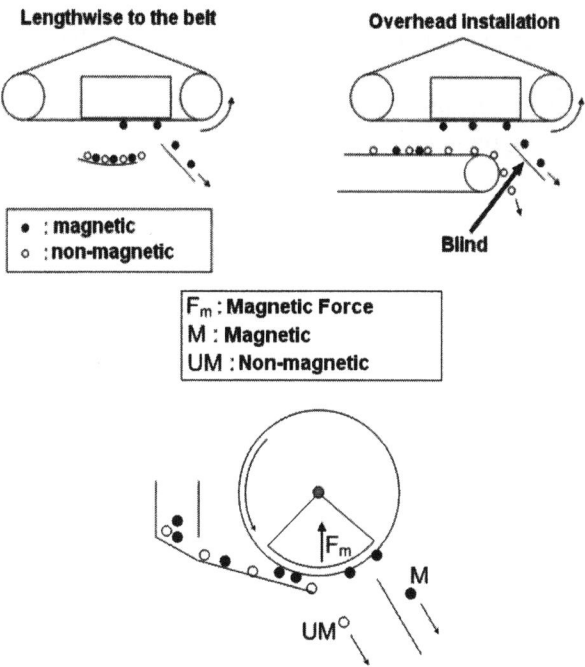

Source: VDI 2009

Depending on what kind of excitation is used it is distinguished between permanent and electro magnets. Furthermore is differentiated between strong and weak magnetic field separators. The magnetic separator can be installed as an overband magnetic separator (lengthwise) over the conveyor belts right above the trajectory of the material or it is installed as a magnetic drum separator or with a magnetic pulley, since small ferrous particles could still remain under a non magnetic layer.

The magnetic material is usually separated from the magnet via blinds (see figure 3). Magnetic separators are used for every type of waste to get ferrous metals that have a positive value on the market and can be substantially recovered in e.g. steel mills.

To apply the magnetic separation unit right behind the crushing unit(s) has an additional reason. Because of the extraction of ferrous metals out of the material flow the usually following non-metal separation units realised by eddy current separators is not limited in its function, because if ferrous metals still were in the material they could cause overheating at the eddy current separator and damage these units.

2.6. Separation of non-ferrous metals

Right behind the ferrous metal separation unit follows the separation of non-ferrous metals (e.g. copper, zinc, etc.) which is realised via eddy current separators. These are divided in excentric and concentric separators what is related to the adjustment of the magnetic pole system which is the most important part of this type of separators.

Figure 4: Non-ferrous (NF) separation

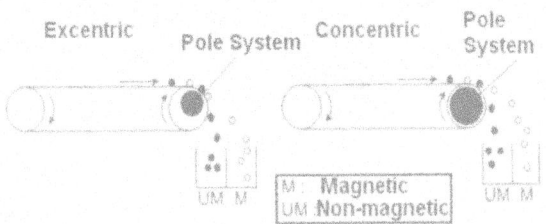

Source: VDI 2009

A *non-ferrous (NF) metal separator* basically consists of a short conveyor driven from the feed end. A rapidly rotating system of permanent magnets – the pole system – which generates high-frequency changing magnetic fields, is incorporated in the head drum (excentric or concentric). These fields create strong eddy currents in the non-ferrous metal parts, in which their own magnetic fields, opposing the external fields, now build up. The NF-metal parts jump out of the remaining material flow. A NF-metal separator can be applied for material with a grain size of 3 to 150 mm.

2.7 All metal separator

In the preparation of solid waste fuel, all-metal separators are mainly applied for plastics processing. High throughputs can be realised if the material is diversified before auto recognition.

Figure 5: All-metal separator

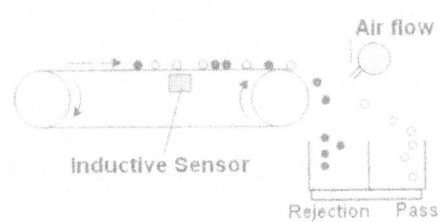

Source: VDI 2009

The previously separated bulk material is transported by a *fast conveyor belt* to a sorting zone. Below the conveyor belt there is an inductive sensor, a detection coil which is placed transverse to the direction of transport and cut into single segments. This sensor analyzes the material over the whole width of the conveyor belt by means of magnetic induction.

As soon as metallic particles are detected, electronic signals are sent to the central computerized control unit. The compressed air jets, individually controlled by the programming, push the detected metals over the diverter gate.

2.8 Near infrared sorting (NIR-sorting)

Material which has to be separated is often fed on a belt conveyor. The conveyor usually operates at fast velocities so that its function is almost like an isolating device.

Figure 6: NIR-sorting

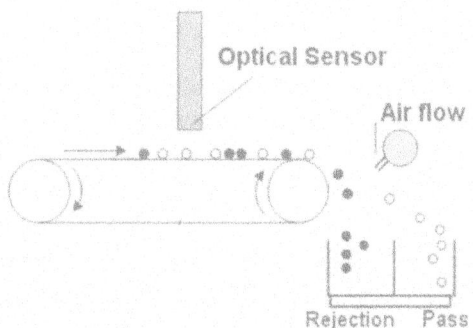

Source: VDI 2009

Halogen lamps and the detector are installed above the belt conveyor. The detector consists of a near infrared spectroscopy (NIR) sensor which scans the whole width of the belt conveyor and transmits the characteristic spectrums of the different materials to a data processor. Signals are then compared with a database.

Sorting then occurs with an air jet batten in front of the discharge end. The air jet lifter is equipped with several single air jets at a distance of about 30 mm apart. Each air jet is fed by a pressure reservoir and is steered by magnetic valves. The data processor transmits a signal if the detection of a particle is positive and the air jet blows it out. Here one or more air jets can be activated. The pressure surge blows out the particle which is then separated from the material flow by a partition plate.

The NIR sorting systems work with a very sensitive spectrometer that detects reflected near infra-red light and evaluates it in line with specific tasks. The NIR "light," which is in the threshold area above visible light, is the result of the reflection of white light on the surface of a material. It is mainly used to separate plastics, glass and wood.

2.9 X-Ray sorting system

Figure 7: X-Ray sorting system

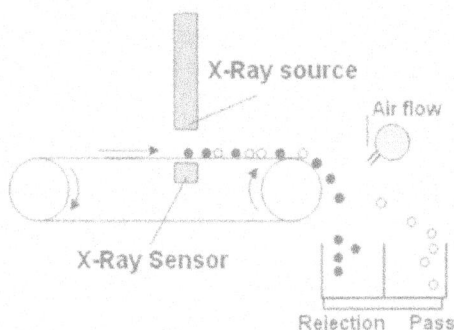

Source: VDI 2009

The technology at the heart of the *X-ray sorting system* (XSS) is based on the transmission of X-rays, i.e. X-rays are shone through the material to be sorted and an X-ray sensitive camera determines the intensity of the transmitted radiation. A computer determines the difference between the incident and the transmitted radiation.

The resulting intensity difference – the absorption – can be used to draw conclusions about the atomic composition of the illuminated material. The system creates an image with almost realistic quality – sufficient to enable image evaluations to be carried out. The absorption depends on both the solid density and the material thickness. The larger the atomic mass and the thicker the material, the greater the absorption. In order to compensate for this technical problem, the material to be sorted is illuminated from two different directions. This is the "Dual-Energy" system. The resulting different transmission paths make it possible to ignore the material thickness.

The XSS X-ray sorting system represents the latest development in the field of sensor-based sorting technology. This technology enables the use of completely new types of sorting criteria for raw materials preparation. The decisive factor for the sorting operation is the density of the material, in particular of its most important chemical elements.

2.10 Heavy media separation

Figure 8: Heavy media separation

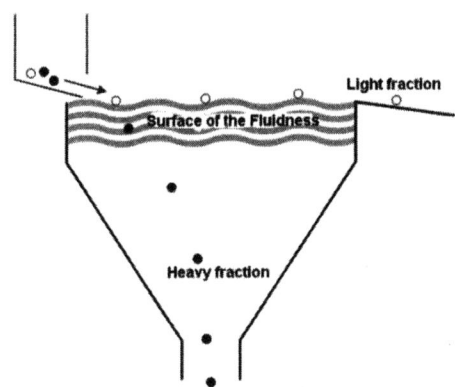

Source: VDI 2009

Heavy media separation is based on different densities of the materials that have to be separated. Material with higher density (heavy fraction) will fall down to the ground and the light fraction swims on the surface of the fluid that serves as the separation medium. The density of the separation fluid must ly between the density of the materials that have to be separated. Separating media can be water, salt solutions (e.g. calcium chlorine) and heavy liquids (suspension of solid matter in water, e.g. kaoline, magnetite and ferro-silicon). To reach good separation results it is necessary that the materials have a unique size of particles and a difference in density of minimum 0,05 g/cm³. Heavy media separation is used for the treatment of non-ferrous metal fractions to separate pure metal fractions and for the separation of plastics.

3. Emissions and consumptions

3.1 Emissions

The figure 9 belong shows input and output flows of a waste treatment plant in general. The emissions must be diversified in direct and indirect emissions of a waste facility.

Figure 9: Consumption and emissions of a waste treatment plant

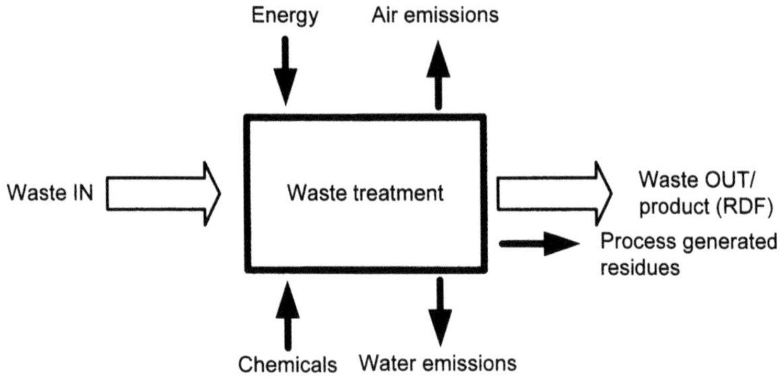

Source: IPPC 2006

For example emissions from energy consumption depend on what kind of energy supply the treatment plant uses. In the field of RDF production treatment plants need electricity (indirect air emissions) and transport (direct air emission). Therefore it is obvious that the type of waste treatment plant determine the amount and kind of emissions of the plant. Subsequent is described what kind of emissions typically occur in the waste treatment sector regarding common waste treatment techniques.

3.2 Generic air emissions from common waste treatments

3.2.1 Volatile Organic Compounds

At present, there are no real data on Volatile Organic Compounds (VOC) available. There are no real data available at present on VOC emissions. The vast majority of sites that undertake air monitoring, undertake it on an irregular basis and are unlikely to take a sample at the times of maximum discharge. Air emissions are particularly difficult to monitor from these sites as operations are generally in the open air and gases are not always controlled.

3.2.2 Fugitive and diffuse emissions

In many installations, fugitive and diffuse emissions may be more significant than point source or channeled emissions. Common examples of the sources are:

- Open vessels (for example, the effluent treatment plant);
- sampling activities;
- storage areas (for example, bays, stockpiles, lagoons, etc.);
- the loading and unloading of containers;
- transferring/bulking up of material from one vessel to another;

- conveyor systems;
- pipe work and ductwork systems (for example, pumps, valves, flanges, catch pots, drains, inspection hatches, etc.);
- poor building containment and extraction;
- potential bypass of abatement equipment (to air or water);
- spillages;
- accidental loss of containment from failed plant and equipment;
- tankers and vessels, manhole openings and other access points;
- displaced vapours in receiving tanks;
- cleaning or replacing of filters;
- drum cutting;
- waste water storage;
- drum storage;
- tank cleaning;
- tanker washing/cleaning.

3.2.3 *Particulate emissions*

Sites handling powders and wastes giving rise to dusts often have particulates to emit to the air.

3.2.4 *Noise and vibration*

'Noise' refers to 'noise and/or vibration' typically detectable beyond the site boundary.

3.2.5 *Odour emissions*

Emissions to air tend only to be checked subjectively by using the sense of smell. Odour emissions are associated with point sources as well as fugitive sources. In addition to ammonia previously discussed, the handling of any substance which is or may contain a VOC (or other odorous substances, for example, mercaptans or other compounds containing sulphur) will potentially lead to odour noticeable in and beyond the installation boundary. Odours may arise from:

- Storage;
- the transfer or bulking up of wastes containing VOCs or other odorous substances;
- a failure to adequately inspect and maintain plant and equipment, which may lead to fugitive emissions, e.g. leaks from pumps.

3.2.6 Generic water emissions from common waste treatments

A distinction can be made between installations conducting 'dry' or solid phase operations, e.g. transfer or stabilisation, which do not produce a distinct liquid effluent; and those conducting liquid phase treatment, e.g. acid neutralisation and oil water separation. 'Dry' processes typically only produce effluents from activities such as from rainwater collection and incidents such as spills and leakages. In general terms, the strength of this effluent in terms of metals and COD levels will be relatively low. 'Wet' processes, in addition to the general effluent arising from yard drainage, etc., produce an effluent from the reaction, precipitation, settlement and dewatering processes. Waste water may be generated in the installations due to:

- Unplanned discharges to drain (e.g. emergency control, fire);
- spillage from storage;
- discharge to storm drain;
- discharge of bund and secondary containment contents;
- process waste water.

Many transfer stations are associated with adjacent treatment plants and all run-off goes into that treatment system where it is treated. Others collect the run-off and tanker this to landfill. Again there is no discharge to receiving waters or sewer. The remainder of the installations discharge either to surface water (unusual option) or to sewer. In the vast majority of EU countries it is not permitted to make direct discharges to sewer or to controlled waters. A security storage is then needed in order to control or treat the water before discharge.

In principle, there will always be small quantities of every material decanted at the site discharged to sewer, due to drips and splashes even if there are no spills recorded. The most common materials to be bulked at transfer stations are dilute acids (often from metal treatment), caustic solutions, oils, non-halogenated solvents and aqueous organic wastes. The discharge is almost certain to contain organic carbons, nitrogen compounds (total nitrogen), chloride, some metals and, when bulking non-halogenated solvents, xylene. Discharges to sewer may reach COD levels of several thousand milligrams per litre. The nature of the discharge depends on the wastes being handled at the installation, which invariably involves a wide variety of substances, thereby resulting in a complex effluent. Emissions to water also occur from washing containers and tanks if this occurs in the waste treatment plant. Liquid discharges may arise from the washing and processing of containers prior to their reprocessing, or from the washing of road tankers. One approach of estimating these emissions is to assume that the residual material in each type of container after emptying is 0,5% of the volume, and that all of this material is washed to sewer. In general, volatile residues from containers of solvent waste are evaporated directly to the air rather than being washed to sewer. General leaks and spills can occur in waste transfer stations. Most sites are on hard standing, and liquid and solid spills are eventually washed away to the main interceptors and then to sewer or to an adjacent treatment plant.

3.2.7 Generic releases to soil and process generated waste from common techniques

Most sites will have a continuous, but small, discharge of waste to the site base-ground due to drips, splashes, crushing residues, pipe connections, oil leaks, etc. and these may be washed to the surface water collection points by rainwater and site cleaning. Tank bottoms are another typical waste when storing waste.

4. Techniques to consider in the determination of Best Available Techniques

The treatment of wastes requires a good knowledge of the incoming and outgoing material of a waste treatment plant. Therefore Best Available Techniques (BAT) is to implement and run a management system to ensure legal compliance and to monitor quality of the input and output (product quality in case of RDF production, secondary raw material production, etc.). Best practise is to implement an environmental management system where all the legal requirements and environmental aspects are assessed and proved regularly.

To monitor materials entering and departing the plant a analysing system is necessary to guarantee compliance with thresholds demanded by waste incineration, RDF incineration and other following treatment plants (IPPC 2006).

5. Best Available Techniques

Best Available Techniques in the waste management sector depends on the targeted product and the required product quality, the type and appearance of the input material and economic factors like market situation for specific materials (e.g. prices for metals, glass, paper and wood.)

Depending on the kind of waste treatment plant BAT is to combine aforementioned techniques (see section 2) in such way that the material recovery rate reaches the maximum and that fractions are nearly free of contraries. It is not possible to define BAT for one special kind of treatment system, because all the techniques target different materials and are used for a wide range of wastes so that only the combination of the techniques can lead to an environmentally and economic benefit.

Management systems like environmental management systems, raw material and utility management systems, knowledge of waste IN and waste OUT and therefore management of waste analysis are considered to be BAT and described in the following sections.

5.1 Environmental management

These are techniques related to the continuous improvement of environmental performance. They provide the framework for ensuring the identification, adoption and adherence to BAT options that nevertheless remain important and can play a role in improving environmental performance of the installation. Indeed, these good housekeeping/management techniques/tools often prevent emissions. A number of environmental management techniques are determined as BAT. The scope (e.g. level of detail) and nature of the Environmental Management System (EMS) (e.g. standardised or non-standardised) will generally be related to the nature, scale and complexity of the installation, and the range of environmental impacts it may have. BAT is to: implement and adhere to an EMS that incorporates, as appropriate to individual circumstances, the following features:

a) Definition of an environmental policy for the installation by top management (commitment of the top management is regarded as a precondition for a successful application of other features of the EMS);
b) planning and establishing the necessary procedures;
c) implementation of the procedures, paying particular attention to structure and responsibility, training, awareness and competence, communication, employee involvement, documentation, efficient process control, maintenance programme, emergency preparedness and response and safeguarding compliance with environmental legislation;
d) checking performance and taking corrective action, paying particular attention to monitoring and measurement;
e) Principles of Monitoring are to observe status and success of corrective and preventive action, maintenance of records, independent (where practicable) internal auditing in order to determine whether or not the environmental management system conforms to planned arrangements and has been properly implemented and maintained;
f) review by top management.

5.1.1 Waste IN (Input material of the facility)

To be able to guarantee a high quality of RDF with nearly continuous values for critical parameters like metal content, chlorine content and calorific value it is compulsory to improve the knowledge of the waste IN, Best Available Technique (BAT) comprises the following five items:

1. *Have a concrete knowledge of the waste IN.* Such knowledge needs to take into account the waste OUT, the treatment to be carried out, the type of waste, the origin of the waste, the procedure under consideration and the risk (related to waste OUT and the treatment).

2. *Implement a pre-receiption procedure containing at least the following items:*
   a) tests for the incoming waste with respect to the planned treatment;
   b) making sure that all necessary information is received on the nature of the process(es) producing the waste, including the variability of the process. The personnel having to deal with the pre-receiption procedure need to be able due to his profession and/or experience to deal with all necessary questions relevant for the treatment of the wastes in the WT facility;
   c) a system for providing and analysing a representative sample(s) of the waste from the production process producing such waste from the current holder;
   d) a system for carefully verifying, if not dealing directly with the waste producer, the information received at the pre-acceptance stage, including the contact details for the waste producer and an appropriate description of the waste regarding its composition and hazardousness;
   e) making sure that the waste code according to the European Waste List (EWL) is provided;
   f) identifying the appropriate treatment for each waste to be received at the installation by identifying a suitable treatment method for each new waste enquiry and having a clear methodology in place to assess the treatment of waste and that considers the individual specifications for the treated waste;

3. *Implement a receiption procedure containing at least the following items:*
   a) a clear and specified system allowing the operator to accept wastes at the receiving plant only if a defined treatment method and disposal/recovery route for the output of the treatment is determined. Regarding the planning for the acceptance, it needs to be guaranteed that the necessary storage, treatment capacity and dispatch conditions (e.g. acceptance criteria of the output by the other installation) are also respected;
   b) measures in place to fully document and deal with acceptable wastes arriving at the site, such as a pre-booking system, to ensure e.g. that sufficient capacity is available;
   c) clear and unambiguous criteria for the rejection of wastes and the reporting of all non conformances;

d) a system for identifying the maximum capacity limit of waste that can be stored at the facility;

e) visually inspect the waste IN to check compliance with the description received during the pre-acceptance procedure. For some liquid and hazardous waste, this BAT is not applicable.

4. *Implement different sampling procedures for all different incoming waste vessels* delivered in bulk and/or containers. These sample procedures may contain the following items:

   a) sampling procedures based on a risk approach. Some elements to consider are the type of waste (e.g. hazardous or non-hazardous) and the knowledge of the customer (e.g. waste producer);

   b) have different sampling procedures for bulk (liquid and solids), large and small containers and laboratory smalls. The number of samples taken should increase with the number of containers. In extreme situations, small containers must all be checked against the accompanying paperwork. The procedure should contain a system for recording the number of samples and degree of consolidation;

   c) details of the sampling of wastes in drums within designated storage, e.g. the timescale after receipt;

   d) sample prior to acceptance;

   e) maintenance of a record at the installation of the sampling regime for each load, together with a record of the justification for the selection of each option there should be a system for determining and recording that include a suitable location for the sampling points, the number of samples and degree of consolidation and the operating conditions at the time of sampling;

   f) a system to ensure that the waste samples are analysed;

   g) in the case of cold ambient temperatures, a temporary storage may be needed in order to allow sampling after defrosting. This may affect the applicability of some of the above items in this BAT.

5. *Have a reception facility covering at least the following issues*:

   a) have a laboratory to analyse all the samples at the speed required by BAT. Typically this requires having a robust quality assurance system, quality control methods and maintaining suitable records for storing the analyses results. Particularly for hazardous wastes, this often means that the laboratory needs to be on-site;

   b) have a dedicated quarantine waste storage area as well as written procedures to manage non-accepted waste. If the inspection or analysis indicates that the wastes fail to meet the acceptance criteria (including, e.g. damaged, corroded or unlabelled drums) then the wastes can be temporarily stored there safely. Such storage and procedures should be designed and managed to promote the rapid management (typically a matter of days or less) to find a solution for that waste;

c) have a clear procedure dealing with wastes where inspection and/or analysis prove that they do not fulfill the acceptance criteria of the plant or do not fit with the waste description received during the pre-acceptance procedure. The procedure should include all measures as required by the permit or national/international legislation to inform competent authorities, to safely store the delivery for any transition period or to reject the waste and send it back to the waste producer or to any other authorized destination;

d) move waste to the storage area only after acceptance of the waste;

e) mark the inspection, unloading and sampling areas on a site plan;

f) have a sealed drainage system;

g) a system to ensure that the installation personnel who are involved in the sampling, checking and analysis procedures are suitably qualified and adequately trained, and that the training is updated on a regular basis;

h) the application of a waste tracking system unique identifier (label/code) to each container at this stage. The identifier will contain at least the date of arrival on-site and the waste code.

5.1.2. *Waste OUT (output material of the facility)*

To assure the production of high quality output materials (e.g. RDF) it is important to improve knowledge of the waste OUT, so that BAT is to analyse the waste OUT according to the relevant parameters important for the receiving facility (landfill, incinerator, etc.).

5.2 Management systems

Different procedures may be needed to take into account the physico-chemical properties of the waste (e.g. liquid, solid), type of waste treatment (WT) process (e.g. continuous, batch) as well as the changes that may occur to the physico-chemical properties of the wastes when the WT is carried out, so that BAT is to:

1. *Have a system in place to guarantee the traceability of waste treatment.* A good traceability system contains the following items:

   a) documenting the treatments by flow charts and mass balances;

   b) carrying out data traceability through several operational steps (e.g. pre-acceptance / acceptance / storage / treatment / dispatch). Records can be made and kept up-to-date on an ongoing basis to reflect deliveries, on-site treatment and dispatches. Records are typically held for a minimum of six months after the waste has been dispatched;

c) recording and referencing the information on waste characteristics and the source of the waste stream, so that it is available at all times. A reference number needs to be given to the waste and needs to be obtainable at any time in the process to enable the operator to identify where a specific waste is in the installation, the length of time it has been there and the proposed or actual treatment route;

d) having a computer database/series of databases, which are regularly backed up. The tracking system operates as a waste inventory/stock control system and includes date of arrival on-site, waste producer details, details on all previous holders, an unique identifier, pre-acceptance and acceptance analysis results, package type and size, intended treatment/disposal route, an accurate record of the nature and quantity of wastes held on-site including all hazards details on where the waste is physically located in relation to a site plan, at which point in the designated disposal route the waste is currently positioned;

e) only moving drums and other mobile containers between different locations (or loaded for removal off site) under instructions from the appropriate manager, ensuring that the waste tracking system is amended to record these changes.

2. *Have and apply mixing/blending rules oriented to restrict the types of wastes* that can be mixed/blended together in order to avoid increasing pollution emission of down-stream waste treatments. These rules need to consider the type of waste (e.g. hazardous, non-hazardous), waste treatment to be applied as well as the following steps that will be carried out to the waste OUT.

3. *Have a segregation and compatibility procedure in place, including:*

e) keeping records of the testing, including any reaction giving rise to safety parameters (increase in temperature, generation of gases or raising of pressure); a record of the operating parameters (viscosity change and separation or precipitation of solids) and any other relevant parameters, such as generation of odours);

f) packing containers of chemicals into separate drums based on their hazard classification. Chemicals which are incompatible (e.g. oxidisers and flammable liquids) should not be stored in the same drum.

4. *Have an approach for improving waste treatment efficiency.* This typically includes the finding of suitable indicators to report WT efficiency and a monitoring programme.

5. *Produce a structured accident management plan.*

6. *Have and properly use an incident diary.*

7. *Have a noise and vibration management plant* in place as part of the EMS. For some WT installations, noise and vibration may not be an environmental problem.

8. *Consider any future decommissioning at the design stage.* For existing installations and where decommissioning problems are identified, put a programme to minimise these problems in place.

5.3 Utilities and raw material management

Every waste treatment facility runs high-tech equipment with important emissions and consumptions such as energy, water and oils, so that Best Available Technique (BAT) is to:

1. Provide for a breakdown of energy supply and generation (including exporting) by the type of source (i.e. electricity, gas, liquid conventional fuels, solid conventional fuels and waste). This involves:
   a) reporting the energy consumption information in terms of delivered energy;
   b) reporting the energy exported from the installation;
   c) providing energy flow information (for example, diagrams or energy balances);
   d) showing how the energy is used throughout the process.

2. *Continuously increase the energy efficiency of the installation, by*:
   a) developing an energy efficiency plan;
   b) using techniques that reduce energy consumption and thereby reduce both direct (heat and emissions from on-site generation) and indirect (emissions from a remote power station) emissions;
   c) defining and calculating the specific energy consumption of the activity (or activities), setting key performance indicators on an annual basis (e.g. MWh/tonne of waste processed) (related to BAT number 1.k and 20).

3. *Carry out an internal benchmarking* (e.g. on an annual basis) of raw materials consumption.

4. *Explore the options for the use of waste as a raw material for the treatment of other wastes.* If waste is used to treat other wastes, then to have a system in place to guarantee that the waste supply is available. If this cannot be guaranteed, a secondary treatment or other raw materials should be in place in order to avoid any unnecessary waiting treatment time.

5.4 Storage and handling

A good management of the stock and the handling of material helps to optimise transportproceses and can help to reduce emissions such as odours and dust, so that BAT is to:

1. *Apply the following techniques related to storage*:
    a) locating storage areas away from watercourses and sensitive perimeters, and in such a way so as to eliminate or minimise the double handling of wastes within the installation;
    b) ensuring that the storage area drainage infrastructure can contain all possible contaminated run-off and that drainage from incompatible wastes cannot come into contact with each other;
    c) using a dedicated area/store which is equipped with all necessary measures related to the specific risk of the wastes for sorting and repackaging laboratory smalls or similar waste. These wastes are sorted according to their hazard classification, with due consideration for any potential incompatibility problems and then repackaged. After that, they are removed to the appropriate storage area;
    d) handling odorous materials in fully enclosed or suitably abated vessels and storing them in enclosed buildings connected to abatement;
    e) ensuring that all connections between the vessels are capable of being closed via valves. Overflow pipes need to be directed to a contained drainage system (i.e. the relevant bunded area or another vessel);
    f) having measures available to prevent building up of sludges higher than a certain level and the emergence of foams that may affect such measures in liquid tanks, e.g. by regularly controlling the tanks, sucking out the sludges for appropriate further treatment and using anti-foaming agents;
    g) equipping tanks and vessels with suitable abatement systems when volatile emissions may be generated, together with level meters and alarms. These systems need to be sufficiently robust (able to work if sludge and foam is present) and regularly maintained.

2. *Separately bund the liquid decanting and storage areas using bunds which are impermeable and resistant to the stored materials.*

3. *Apply the following techniques concerning tank and process pipe work labelling*:
    a) clearly labelling all vessels with regard to their contents and capacity, and applying an unique identifier. Tanks need to have an appropriately labeled system depending on their use and contents;

b) ensuring that the label differentiates between waste water and process water, combustible liquid and combustible vapour and the direction of flow (i.e. in or outflow);
c) keeping records for all tanks, detailing the unique identifier; capacity; its construction, including materials; maintenance schedules and inspection results; fittings; and the waste types which may be stored/treated in the vessel, including flashpoint limits.

4. *Take measures to avoid problems that may be generated from the storage/accumulation of waste.*

5. *Apply the following techniques when handling waste:*
    a) having systems and procedures in place to ensure that wastes are transferred to the appropriate storage safely;
    b) having in place a management system for the loading and unloading of waste in the installation, which also takes into consideration any risks that these activities may incur. Some options for this include ticketing systems, supervision by site staff, keys or color-coded points/hoses or fittings of a specific size;
    c) ensuring that a qualified person attends the waste holder site to check the laboratory smalls, the old original waste, waste from an unclear origin or undefined waste (especially if drummed), to classify the substances accordingly and to package into specific containers. In some cases, the individual packages may need to be protected from mechanical damage in the drum with fillers adapted to the packaged waste properties;
    d) ensuring that damaged hoses, valves and connections are not used;
    e) collecting the exhaust gas from vessels and tanks when handling liquid waste;
    f) unloading solids and sludge in closed areas which are fitted with extractive vent systems linked to abatement equipment when the handled waste can potentially generate emission to air (e.g. odours, dust, VOCs);
    g) using a system to ensure the bulking of different batches only takes place with compatibility testing.

6. *Ensure that the bulking/mixing to or from packaged waste only takes place under instruction and supervision and is carried out by trained personnel. For certain types of wastes, such as bulking/mixing needs to be carried out under local exhaust ventilation.*

7. *Ensure that chemical incompatibilities guide the segregation required during storage.*

5.5 Air emission treatments

To prevent or control the emissions mainly of dust, odours and VOC and some inorganic compounds, Best Available Technique (BAT) is to:

1. *Restrict the use of open topped tanks, vessels and pits by:*
   a) not allowing direct venting or discharges to air by linking all the vents to suitable abatement systems when storing materials that can generate emissions to the air (e.g. odours, dust, VOCs);
   b) keeping the waste or raw materials under cover or in waterproof packaging;
   c) connecting the head space above the settlement tanks (e.g. where oil treatment is a pretreatment process within a chemical treatment plant) to the overall site exhaust and scrubber units.

2. *Use an enclosed system with extraction*, or under depression, to a suitable abatement plant. This technique is especially relevant to processes which involve the transfer of volatile liquids, including during tanker charging/discharging.

3. *Apply a suitably sized extraction system* which can cover the holding tanks, pretreatment areas, storage tanks, mixing/reaction tanks and the filter press areas, or to have in place a separate system to treat the vent gases from specific tanks (for example, activated carbon filters from tanks holding waste contaminated with solvents) (see Section 4.6.1).

4. *Correctly operate and maintain the abatement equipment*, including the handling and treatment/disposal of spent scrubber media.

5. *Have a scrubber system in place for the major inorganic gaseous releases* from those unit operations which have a point discharge for process emissions. Install a secondary scrubber unit to certain pretreatment systems if the discharge is incompatible, or too concentrated for the main scrubbers.

6. *Have leak detection and repair procedures in place* in installations a) handling a large number of piping components and storage and b) compounds that may leak easily and create an environmental problem (e.g. fugitive emissions, soil contamination) (see Section 4.6.2). This may be seen as an element of the EMS (see BAT number 1).

7. *Reduce air emission to the following levels* (see by using a suitable combination of preventive and/or abatement techniques. The techniques available on the market fulfill these values.

Table 2: Emission levels associated to the use of BAT

Air parameter	Emission levels associated to the use of BAT (mg/Nm^3)
VOC	7-20^a
PM	5-20

^{a} *For low VOC loads, the higher end of the range can be extended to 50*
Source: IPPC 2006

6. Emerging Techniques

Emerging technique is understood in this document as a novel technique that has not yet been applied in any industrial sector on a commercial basis. This chapter contains those techniques that may appear in the near future and that may be applicable to the waste treatment sector.

6.1 On-line analysis

The technique of the online-analysis is one of the latest developments on the field of analysis and quality assurance. It can be used for all applications in the field of preparation of solid recovered fuels. On-line analysis is used for crushed and/or for non-crushed materials with automatically elimination of materials which do not comply with the quality criteria for e.g. solid recovered fuels – especially when the chlorine- and/or bromine values are exceeded.

The mode of function is based on a new X-ray fluorescence-analysis with high speed analysis, so that a large quantity of crushed or not crushed materials (it depends on technical performance and determination) per hour can be analysed and/or detected and it can be automatically eliminated by overdraw nominal stock. The configuration of the measuring unit and/or analyser takes place directly above a conveyor. A material stream as uniform as possible is directed under the measuring-unit and/or analyser and is analysed and/or measured. If a limit-value is exceeded an electronically signal (digital or analogue) follows. There upon controlled through a software and/or electronics-unit the objectionable material is automatically (mechanically, hydraulically, pneumatically, electrostatically or magnetically) discharged. The measuring-unit and/or analyser can be equipped with one or more X-ray tubes or with one or more detectors. As a additional control and quality assurance for the material input, also a handheld-unit can be used.

The handheld-unit is also based on the X-ray fluorescence-method and it can especially be used for analysis and/or detection of chlorine, bromine and heavy metals. The following elements can be analysed and detected with this tool (depending on equipment and software): Cl, Br, Cd, Hg, Pb, As, Se, Ni, Sb, Cu, Ba,

Cr, Sn, Mo, Zn, Sr, Fe, Co, Ti, V, Rb, Ir, Pt, Au, Ag, Pd, Nb, W, Bi, Mn, Ta, Zr, Hf, Re.

6.2 Hazardous waste preparation for energy recovery

New adsorbents for the preparation of solid waste fuel from hazardous waste. There is a permanent research for other absorbents in order to replace the fresh sawdust.

6.3 Preparation of solid fuel from organic/water mixtures

The process consists in preparation of a fuel for the use in cement kilns. The process is the mixing of the organic-water mixtures with a lime hydrate porous structure in order to capture the organics and use such product as raw material in the cement industry. This technique is able to deal with clinical waste, municipal waste, hazardous/chemical waste and non-hazardous industrial and commercial waste.

7. Concluding remarks

RDF production from wastes has a high demand on quality of the RDF and this means that treatment plants that produce RDF from solid wastes need an excellent processing and monitoring management system for their materials and a well selected and connected system of sorting aggregates (see chapter 2) to achieve the high level of thresholds requested by incineration and co-incineration plants. High quality RDF for example with low heavy metal and chlorine content could easily be co-incinerated in cement kilns or in steel works which have the highest level of requirements. To reach these levels the composition of the Waste IN is the limiting factor. It is proved that RDF from municipal and household wastes can achieve these thresholds. On the other hand is the usage of industrial wastes associated with a much higher effort to achieve these limits so that RDF from industrial wastes is mainly produced for co-incineration in municipal waste incineration plants or for incineration in special RDF incineration plants that can technically handle a RDF that contains some heavy metals, chlorine and where the calorific value is oscillating around 15.000 kJ/kg.

Against the background of constriction of raw materials and natural resources like oil, gas and coal, waste becomes a more and more important secondary energy resource that helps to unburden our environment and delivers heat and electricity. A waste industry where wastes are processed in a cycle moves closer by applying the Best Available Techniques.

8. References

(IPPC 2006) European Commission, Integrated Pollution Prevention and Control, "Reference Document on Best Available Techniques for the Waste Treatment Industries", August 2006

(TCA 2007) Turkish Court of Accounts, "Waste Management in Turkey, National regulations and implementation results", Performance Audit Report, January 2007

(VDI 2009) VDI 2342, Verein Deutscher Ingenieure (Association of german engineers), draft of the guideline for treatment of electronical devices (No. 2342), Jan. 2009

Annex

Expertise Catalogue

Kerstin Kuchta, Konstantin Haker, Georgi Chobankov

1. General information of renewables in Turkey

1.1 Potential for RE in Turkey: Sources of References

General Directorate of Electrical Power Resources Survey and Development Administration (EIE). Renewable energy resources, solar energy, solar energy studies, http://www.eie.gov.tr/turkce/gunes/eiegunes.html, 2006.

http://cat.inist.fr/?aModele=afficheN&cpsidt=20053566.

http://emeraldinsight.com/Insight/ViewContentServlet;jsessionid=59946E1A6A837378677E0CC29B467EC3?Filename=Published/EmeraldFullTextArticle/ Articles/083017050 2.html.

http://traccess.tubitak.gov.tr/fp6_yeni/DefaultIframe_en.aspx?aId=529.

http://turkey-electricity.com/page12.html.

http://www.dsi.gov.tr/english/congress2007/chapter_2/27.pdf

http://www.dsi.gov.tr/english/congress2007/chapter_2/57.pdf

http://www.iea-pvps.org/ar/ar07/07ar_Turkey.pdf

http://www.mondaq.com/article.asp?articleid=59066

http://www.sciencedirect.com/science?_ob= ArticleURL&_udi= B6V4S-4S4S5 JW-1&_user= 10&_coverDate= 11%2F30%2F2008&_alid= 799492189&_rdoc = 1&_fmt= high&_orig= search&_cdi= 5766&_sort= d&_docanchor= &view= c&_ct= 7&_acct= C000050221&_version= 1&_urlVersion= 0&_userid= 10&md5= a8dee198ad50ff5fa57c74ffcb7178e1

http://www.sciencedirect.com/science?_ob= MiamiImageURL&_imagekey= B6V 2P-458N8PC-2-8&_cdi= 5708&_user= 4478132&_check= y&_orig= search&_ coverDate= 02%2F28%2F2003&view= c&wchp= dGLbVzz-zSkWb&_valck= 1&md5= 89757551f86e8fda48baecfb1f05fccf&ie= /sdarticle.pdf

http://www.thetravelfoundation.org.uk/assets/tools_training_guidelines/policy%20papers/turkey.pdf

http://www.turkey-electricity.com/page6.html

http://www.turkey-electricity.com/page9.html

http://www.turkey-electricity.com/page15.html

http://www.windfair.net/press/5130.html

International Energy Agency (IEA) (2007)

Ozdamar, A., Gursel, K.T., Orer, G. and Pekbey, Y. (2004). Investigation of the potential of wind–waves as a renewable energy sources: by the example of Cesme, Turkey, Renew Sustain Energy Rev 8 (2004), pp. 581–592

Turkey Electricity-Renewable Energy, http://turkey-electricity.com/page9.html

Turkish Weekly, The Investment Potential of the Turkish Energy Market, Fevzi Saffet Bora (04 February 2007)

1.2 RE Policy in Turkey and EU: Source References

Bilgen, Selçuk, Keleş, Sedat, Kaygusuz, Abdullah, Sarı, Ahmet and Kaygusuz, Kamil (2006). Global warming and renewable energy sources for sustainable development: A case study in Turkey.

http://ec.europa.eu/energy/res/legislation/share_res_eu_en.htm

http://www.balkanlight.eu/abstracts_pdf/b13.pdf

http://www.energy.eu/renewables/factsheets/2008_res_sheet_sweden_en.pdf

http://www.erec.org/fileadmin/erec_docs/Projcet_Documents/RES2020/SWEDEN_RES_Policy_Review_April_2008.pdf

http://www.erneuerbare-energien.de/inhalt/print/4306/

http://www.euractiv.com/en/energy/eu-renewable-energy-policy/article-117536

http://www.europarl.europa.eu/meetdocs/2004_2009/documents/fd/d-tr20060425_06/d-tr20060425_06en.pdf

http://www.eva.ac.at/enercee/enlargement.htm

http://www.hightech-strategie.de/en/36.php

http://www.iea.org/textbase/pm/?mode=re&id=2547&action=detail

http://www.invest.gov.tr/documents/publications/turkey2005.pdf

http://www.planbleu.org/publications/atelier_energie/TR_National_Study_Final.pdf

http://www.planbleu.org/publications/atelier_energie/TR_Summary.pdf

http://www.rec.org/REEEP/energy_country_profiles/turkey.pdf

http://www.regeringen.se/content/1/c6/06/47/22/2c000830.pdf

http://www.regeringen.se/sb/d/5745/a/19594

http://www.sgu.se/sgu/eng/samhalle/energi-klimat/fornybar-energi_info_e.html

http://www.sweden.gov.se/sb/d/5745/a/19594

http://www.wealthdaily.com/articles/renewable-energy-germany/1418

Jacobsson, S. and Lauber, V. (in press). The politics and policy of energy system transformation – explaining the German diffusion of renewable energy technology. *Energy / Policy*.

Law on Utilization of Renewable Energy Resources for the Purpose of Generating Electrical Energy, Law No. 5346, Ratification Date: 10.05.2005, Enactment Date: 18.05.2005. SECTION ONE. Purpose, Scope, Definitions and Abbreviations

2. Working Papers

2.1. Biogas: References

Biogashandbuch Bayern, edited by Landesamt für Umwelt und Naturschutz Bayern, Germany (15 July 2007)

Bundesforschungsanstalt für Landwirtschaft (Federal Research Institute for agriculture) (2007). Forum bioenergy villages, Göttingen, March 2007

Demirel, Burak (2009). "The potential and opportunities of biogas use in Turkey", Bogazici University, Institute of Environmental Sciences, February 2009

Deutsche BiomasseForschungsZentrum gemeinnützig GmbH, "Stand der Biogastechnik" (Entwurf vom 15.10.2008)

Fachverband Biogas e.V. (German biogas Association) (February 2009)

German Federal Ministry for Economy and Work (2007). "Technical basics for the assessment of biogas plants"

Kizilaslan, N. & Kizilaslan, H. (2007). Department of Agricultural Economics, Faculty of Agriculture, Gaziosmanpasa University, Turkey; "Energy Sources", Part B: Economics, Planning, and Policy, Vol. 2, Issue 3 July 2007, pp. 277-286

Ministry of Environment and Forest, German Federal State Rheinland-Pfalz (2009). Manual on the construction and operation of biogas plant, February 2009

Petersson, Anneli (2009). "Biogas from international perspective", Swedish Gas Centre, February 2009

Reinhold, Günther (2007). „Biogas – Technik und Trends", Thüringer Landesanstalt für Landwirtschaft, Pressemitteilung Nr. 40, September 2007

Weiland, P., Rieger, Ch. & Ehrmann, Th. (2003). „Evaluation of the newest biogas plants in Germany with respect to renewable energy production, greenhouse gas reduction and nutrient management", Institute of Technology and Biosystems Engineering Federal Agricultural Research Centre (FAL), Future of Biogas in Europe II, Esbjerg 2-4 October 2003

2.1.1 Further Information:

National Associations (selection):

Alphéeis:
http://www.alpheeis.com

Association Eden:
http://www.eden-enr.org

Association Hespul:
http://www.hespul.org

Austrian Biomass Association:
http://www.biomasseverband.at

Biogas Associatio:
http://www.biogas.org.uk

Biogas Association of Denmark:
http://www.biogasbranchen.dk

Biogasunion e.V:
http://www.biogasunion.de/

British BioGen:
http://www.britishbiogen.co.uk

Cheshire Renewable Energy Initiative Cheshire County Council, Environmental Planning:
http://www.cheshirerenewables.org.uk

CLER – Comité de Liaison Energies Renouvelables:
http://www.cler.org

Cnam IFFI – Institut Français du Froid Industriel et du génie climatique:
http://www.cnam.fr/institut/iffi

CZ BIOM – Czech Biomass Association:
http://www.vurv.cz

Danish Biogas Plant Association:
http://www.biogasdk.dk/ffdb_ENG.htm

De Verband Group:
http://www.de-verband.com/portal/index.html

Energieagentur Judenburg-Knittelfeld-Murau:
http://www.energieagentur.ainet.at

Energy Agency of Satakunta:
http://www.prizz.fi

Energy and Environmental Office of Viborg:
http://www.biogasinfo.org

Enver Enerji Ve Cevre Ltd:
http://www.enver.com.tr

HBA – Hungarian Biomass Association:
http://www.mbmt.hu

IOB – Institute of Biology, UK:
http://www.iob.org

NL-BEA – Netherlands Bio-energy Association:
http://www.xs4all.nl

NOBIO – Norwegian Bioenergy Association:
http://www.nobio.no

Österreichischer Biomasse-Verband:
http://www.biomasseverband.at/biomasse

OVE – The Danish Organisation for Renewable Energy:
http://www.orgve.dk

Renewable Power Association:
http://www.r-p-a.org.uk

SBGF – Swedish Biogas Association:
http://www.sbgf.org

SK-BIOM – Slovak Biomass Association:
http://www.skbiom.sk

Swiss Biogas Forum – Links:
http://www.biogas.ch/links.htm

Syndicat intercommunal du Grand Thessaloniki:
http://www.hyper.gr/asstota

The Finnish Biogas Association- Links:
http://www.kolumbus.fi/suomen.biokaasukeskus/en/enindex.htm

ValBiom – Valorization of Biomass:
http://www.valbiom.be

International Associations (selection)

Biogas Accepted:
http://www.biogasaccepted.eu/index.php?id=6

ENGVA – European Natural Gas Vehicle Association:
http://www.engva.org

European Algae Biomass Association:
http://www.eaba-association.eu/

European project for development of biogas in transports:
http://www.biogasmax.eu/links-to-biogas-project-partners/links-to-biogas-and-biofuel-project-partners.html

European Renewable Energy Research Centres Agency:
http://www.eurec.be/

International Renewable Energy Agency:
http://www.irena.org/

RAMIRAN – Links:
http://www.ramiran.net/index.php?page=links

The European Biomass Association:
http://www.aebiom.org/

Companies/Industry (selection)

http://www.biogas-nord.com

http://www.biogas-weser-ems.de

http://www.brgbiogas.com/

http://www.farmatic.de

http://www.haase-energietechnik.de

http://www.oekobit-biogas.com

http://www.scandinavianbiogas.se

http://www.schmack-biogas.com/

http://www.sgc.se

http://www.swedishbiogas.eu

http://www.tianren.com

http://www.uts-italia.it/

Universities (selection)

HAW Hamburg, CC4E – Competence Center für Erneuerbare Energien und Energieeffizienz:
http://www.haw-hamburg.de/energie.html

Kettering University:
http://www.kettering.edu

Loughborough *University, The Renewable Energy Centre*:
http://www.therenewableenergycentre.co.uk/

Mälardalen University, RG „Process & sensor development":
http://www.mdh.se/hst/forskning/PRO/research_groups/process_sensor_development

Swedish University of Agricultural Sciences, BTC-Biofuel Technology Centre:
http://www.btk.slu.se/ShowPage.cfm?OrgenhetSida_ID=7675

University of North Dakota, Energy & Environmental Research Center (EERC):
http://www.undeerc.org/

Portals (selection):

Biomass Info Centre:
http://www.bio-energie.de/

Branchen Domain – Links Biogas:
http://www.branchen-domain.de/branchen/branchenverzeichnis/branche-516_biogasanlagen.htm

C.A.R.M.E.N.:
http://www.carmen-ev.de/

Checkbiotech:
http://bioenergy.checkbiotech.org/

Cheshire Renewable Energy Initiative:
http://www.cheshirerenewables.org.uk

Dena:
http://www.renewables-made-igermany.com/de/biogas/

EUBIONET – European Energy Networks:
http://eubionet.vtt.fi/

Fachagentur Nachwachsende Rohstoffe e. V. (FNR):
http://www.nachwachsende-rohstoffe.de/

Online Planung:
http://www.biogas-netzeinspeisung.at/

Search Engine for Renewable Energy and Energy Efficiency:
http://www.reegle.info/

The Bioenergy Site:
http://www.thebioenergysite.com/

The European Anaerobic Digestion Network:
http://www.adnett.org

2.1.2 Biogas in the EU: References

Biogas Barometer (2008). EurObserv'ER.

DaSilva, E.J. (1979). Biogas generation: developments. problems, and tasks – an overview. In *Bioconversion of Organic Residues for Rural Communities*. Tokyo: United Nations University Press.

Fischer, T. & Krieg, A. (2002, August 2). *Krieg & Fischer Ingenieure GmbH*. Retrieved December 9, 2008, from www.kriegfischer.de

Helmut Kaiser Consultancy (2008, July). Retrieved December 1, 2008, from http://www.hkc22.com/biogas.html

Hofmann, F. (2007). *Biogasanlagen – Nachhaltiger Beitrag zum Klima- und Umweltschutz*. Leipzig.

Kampschulte, P.D. (2007). *Energieerzeugung aus Biomasse*. Hamburg, Germany.

Krieg, A. & Fischer, T. (2002). *Biogas-Anlagentechnik – speziell Vergärung von NaWaRos*. Göttingen.

Lindblom, P. (2008, March 21). *Kristianstad kommun*. Retrieved December 16, 2008, from Saubere Abgase, wenn Biogas genutzt wird: http://www.kristianstad.se/sv/Kristianstads-kommun/Sprak/Deutsch/Umwelt/Biogas/Treibstoff- der-Zukunft/Emissionen/

Petersson, A. (2008). *Biogas from an international perspective*. Swedish Gas Centre.

Rademakers, L. & Van den Berg, J. (2008, February 15). *Biopact*. Retrieved December 1, 2008, from http://biopact.com/2008/02/german-group-invests-50-million-in-20.html

– (2008, February 16). *Biopact*. Retrieved December 1, 2008, from http://biopact.com/2008/02/branson-sees-future-in-african-biofuels.html

Rutz M.Sc., D.D. & Janssen, D.R. (2008). Support for the biogas market in Southern and Eastern Europe. *forum.newpower*, 20-21.

Scandinavian Biogas (2008). Retrieved December 14, 2008, from Why Biogas: http://www.scandinavianbiogas.se/web/_biogas_market.html

van Thuijl, E., Roos, C.J. & Beurskens, L.W. (2008). *An Overview of Biofuel Technologies, Markets and Policies in Europe*. Amsterdam.

Werner, U., Stohr, U. & Hees, N. (1989). *Biogas plants in animal husbandry*. GTZ.

Wikipedia (2008, December 5). Retrieved December 15, 2008, from http://en.wikipedia.org/wiki/Biogas

2.2 Photovoltaic: References

BMU publication erneuerbare Energien in Zahlen – nationale und international Entwicklung" KI III (Stands Juli 2008)

European Union in Directive 96/61/EC of 24 September 1996 concerning integrated pollution prevention and control Electricity production from renewable sources, technical potential of RES and electricity generation costs, Office for Official Publications of the European Communities, European Communities, 2005

European Commission, Directorate general jrc Institute for prospective technological studies, sustainability in industry, energy and transport, IPPC bureau. (From BREF document)

http://eippcb.jrc.ec.europa.eu/

http://wohnen.pege.org/2007-intersolar/photovoltaik-co2.htm

http://www.dgs-berlin.de/fileadmin/PDF/PV3_D4UT.pdf

http://www.ise.fhg.de/veroeffentlichungen/nach-jahrgaengen/2004/multicrystalline-silicon-solar-cells-exceeding-20-efficiency

ISBN 92-894-8004-1 © European Communities, 2005

Luxembourg: Office for Official Publications of the European Communities, 2005

Photon international 03/2006

www.heise.de

www.pvresourcrs.com

www.solarserver.de/wissen/photovoltaik

www.tms.org

2.2.1 Tables

25 Largest Solar power plants worldwide:
www.Pvresources.com

Cell packaging *PV3_D4UT_ZellenModule2008*:
http://www.dgs-berlin.de/fileadmin/PDF/PV3_D4UT.pdf

Cell-types:
http://www.dgs-berlin.de/fileadmin/PDF/PV3_D4UT.pdf

Effect of Temperature:
www.Photon.de

Examples for cell packaging (Ethylene-Vinyl-Acetate (EVA)):
PV3_D4UT_ZellenModule2008

Percentage of renewable Energy from total Energy consumption BMU publication "Erneuerbare Energien in Zahlen-nationale und international Entwicklung" KI III, Stands July 2008:
http://www.bmu.de/english/current_press_ releases/pm/42210.php

Public Expenditure on PV research and market deployment in 2002:
http://ec.europa.eu/research/energy/pdf/vision-report-final.pdf

Regulatory framework for PV in EU-25 and Switzerland (2004):
http://ec.europa.eu/research/energy/pdf/vision-report-final.pdf

Summary of cell qualities:
http://www.dgs-berlin.de/fileadmin/PDF/PV3_D4UT.pdf

2.2.2 Figures

Altitude, Azimuth sun tracker. „Ertragspotenzial nachgeführter Photovoltaik in Europa: Anspruch und Wirklichkeit", www.zsw-bw.de

Amorphous Silicon:
http://www.dgs-berlin.de/fileadmin/PDF/PV3_D4UT.pdf

Band Gaps in eV:
http://solarserver.de

Cadmium-Telluride (CdTe):
http://www.dgs-berlin.de/fileadmin/PDF/PV3_D4UT.pdf

CIS glass cell:
PV3_D4UT_ZellenModule2008

Concentrator cells (II-V Semiconductor):
http://www.tms.org

Copper-Indium-di-Selenide:
http://www.dgs-berlin.de/fileadmin/PDF/PV3_D4UT.pdf

CSG crystalline silicon on glass:
PV3_D4UT_ZellenModule2008

Distribution of cell production by technology Luxembourg: Office for Official Publications of the European Communities, 2005 ISBN 92-894-8004-1 © European Communities, 2005

EFG processed polycrystalline silicon:
http://www.gtsolar.com

High performance silicon cells:
http://www.igafa.de

Horizontal axle sun trackers. „Ertragspotenzial nachgeführter Photovoltaik in Europa: Anspruch und Wirklichkeit", www.zsw-bw.de

Microcrystalline silicon solar. SEM picture of a microcrystalline Si:H film. Source: Applied Films

Micromorph silicon solar cell:
http://www.dgs-berlin.de/fileadmin/PDF/PV3_D4UT.pdf

Polar mounting:
„Ertragspotenzial nachgeführter Photovoltaik in Europa: Anspruch und Wirklichkeit", www.zsw-bw.de

Radiation in Europe. Source: Photon international 03/2006

Sliver solar cells technology:
http://spie.org/Images/Graphics/Newsroom/Imported-2009/1593/1593_fig2.jpg

Spheral solar cells ball:
http://www.technologyreview.com

Standard silicon p-doped:
http://www.svmi.com

String Ribbon Process:
http://www.store.altenergystore.com

Total PV installed power in selected European countries by the End 2003:
http://technologies.ew.eea.europa.eu/News/new853856/

Vertical axle. „Ertragspotenzial nachgeführter Photovoltaik in Europa: Anspruch und Wirklichkeit", www.zsw-bw.de

Wafer based crystalline silicon cells:
www.solarserver.de

Worlds Radiation:
http://www.pvresourcrs.com

2.2.3 Further Information

National Associations (selection)

Arbeitsgemeinschaft Erneuerbare Energie:
http://www.aee.at/

Association Hespul:
http://www.hespul.org

British Photovoltaic Association:
http://www.pv-uk.org.uk

Bundesverband Solare Mobilität:
http://www.solarmobil.net/

Bürger-Solarkraftwerke Rosengarten:
http://buergersolarkraftwerke-rosengarten.de/

Deutsche Gesellschaft für Sonnenenergie:
http://www.dgs.de/

Klima: Aktiv Solarwärme:
http://www.solarwaerme.at

OVE – The Danish Organisation for Renewable Energy:
http://www.orgve.dk

Photovoltaic Austria Federal Association:
http://www.bv-pv.at

Photovoltaic Austria Federal Association:
http://www.bv-pv.at/content/default.asp

Photovoltaik-Verband:
http://www.photovoltaik-verband.de/Home.html

Regio Solar:
http://www.regiosolar.de/startseite/

Renewable Power Association:
http://www.r-p-a.org.uk

Solarenergie-Förderverein Deutschland e.V. (SFV):
http://www.pv-ertraege.de/

Stuttgart SOLAR:
http://www.stuttgart-solar.de

Swiss Association for Solar Energy:
http://www.swissolar.ch

International Associations (selection)

European Photovoltaic Industry Association:
http://www.epia.org/

EUROSOLAR – European Solar Energy Association:
http://www.eurosolar.de/

International Photovoltaic Equipment Association:
http://www.ipvea.com/

International Solar Energy Society:
http://www.ises.org

SEREF, Solar Energy Research and Education Foundation:
http://www.seref.org/

Solar Electric Power Association:
http://www.solarelectricpower.org/

Solar Energy International:
http://www.solarenergy.org

Companies/ Industry (selection)

http://www.conergy.de

http://www.euu.ch/

http://www.firstsolar.com/

http://www.gpv-solar.com/

http://www.hareonsolar.com

http://www.hei.at/

http://www.intersolar.gr

http://www.latitudesolar.com/

http://www.mappabioedilizia.it

http://www.pvesweden.se

http://www.solarline.ch

http://www.solarpower.cz

http://www.solartec.cz

http://www.solartec.gr

http://www.solarvolta.nl

http://www.solems.com

http://www.solosolrenovables.com

http://www.sunergsolar.com

http://www.suninteractiv.de

http://www.sunradiant.it

http://www.suntechnics.com/

http://www.thermosol.it

Universities (selection)

EWE-Forschungszentrum für Energietechnologie:
http://www.ewe-next-energy.de

HAW Hamburg, CC4E – Competence Center für Erneuerbare Energien und Energieeffizienz:
http://www.haw-hamburg.de/energie.html

Loughborough University, The Renewable Energy Centre:
http://www.therenewableenergycentre.co.uk/

RWTH Aachen University, E.ON Energy Research Center:
http://www.eonerc.rwth-aachen.de

TU Wien, EEG- Energy Economics Group:
http://www.eeg.tuwien.ac.at/

Uni Würzburg, Das Bayerische Zentrum für Angewandte Energieforschung:
http://www.zae.uni-wuerzburg.de/

University of Notre Dame, Energy centre:
http://energycenter.nd.edu/

Uppsala Universitet, Ångström Solar Center:
http://www.asc.angstrom.uu.se/

Portals (selection)

Agentur für Erneuerbare Energien:
http://www.unendlich-viel-energie.de/

Arbeitskreis Thermische Solartechnik & Photovoltaik:
http://www.sonnenernte.net/

Cheshire Renewable Energy Initiative:
http://www.cheshirerenewables.org.uk

deSOLaSOL- European co-operation project:
http://www.desolasol.org/en/index/

European Directory for Renewable Energy Companies:
http://www.solarpages.eu/

Global Approval Program For Photovoltaics:
http://www.pvgap.org/

Info site about solar energy:
http://www.pvresources.com

Natural & Renewable Energy Sources:
http://www.clean-energy-ideas.com

Project Soleil Marguerite:
http://www.soleilmarguerite.org/

Search Engine for Renewable Energy and Energy Efficiency:
http://www.reegle.info/

Solar Server, Vereine:
http://www.solarserver.de/initiative/vereine.html

Zeitschrift Photovoltaik:
http://www.photovoltaik.eu/

2.3 Waste to Energy: References

European Commission, Integrated Pollution Prevention and Control, "Reference Document on Best Available Techniques for the Waste Treatment Industries", August 2006

Turkish Court of Accounts, "Waste Management in Turkey, National regulations and implementation results", Performance Audit Report, January 2007

2.3.1 Further information

National Associations (selection)

B9 Organic Energy:
http://www.b9organicenergy.co.uk/

VBSA-Links:
http://www.vbsa.ch/index.html?&page_id=20&node=18&level=0&l=2

Waste to Energy:
http://www.wastetoenergy.co.uk/

International Associations (selection)

Confederation of European Waste-to-Energy Plants:
http://www.cewep.eu

Energy from *Waste Association* (EWA):
www.energy.rochester.edu/uk/ewa/

Companies/ Industry (selection)

http://www.bta-technologie.de

http://www.energ.co.uk

http://www.eon-energyfromwaste.com/

http://www.fromwastetoenergy.com

http://www.kompogas.com

http://www.nehlsen.com/index.php?id=106

http://www.organic-power.co.uk

http://www.polymerenergy.com

http://www.scandinavianbiogas.se

http://www.wastetoenergy.net/

http://zorg-biogas.com/

Universities (selection)

HAW Hamburg, CC4E – Competence Center für Erneuerbare Energien und Energieeffizienz:
http://www.haw-hamburg.de/energie.html

Mälardalen University, RG „Process & sensor development":
http://www.mdh.se/hst/forskning/PRO/research_groups/process_sensor_development

University of Glamorgan, Sustainable Environment Research Centre (SERC):
http://serc.research.glam.ac.uk

Portals (selection)

Anaerobic Digestion:
http://www.anaerobic-digestion.com/

Energy Recovery Council:
http://www.wte.org

The Bioenergy Site:
http://www.thebioenergysite.com/

Waste to Energy Expo:
http://www.wte-expo.com/

WTERT-Links:
http://www.seas.columbia.edu/earth/wtert/olinks.html
http://www.wasteintoenergy.org/

2.4. Waste Management: References

European Commission, Integrated Pollution Prevention and Control, "Reference Document on Best Available Techniques for the Waste Treatment Industries", August 2006

Turkish Court of Accounts, "Waste Management in Turkey, National regulations and implementation results", Performance Audit Report, January 2007

VDI 2342, Verein Deutscher Ingenieure (Association of german engineers), draft of the guideline for treatment of electronical devices (No. 2342), Jan. 2009

Further Information

National Associations (selection)

Association of Local Authorities of Greater Thessaloniki Area, GREECE:
http://www.hyper.gr/asstota

Bundesverband der Deutschen Entsorgungswirtschaft:
http://www.bde.org

Dutch Waste Management Association:
http://www.verenigingafvalbedrijven.nl/

Entsorgungsverband Saar:
http://www.evs.de

Entsorgungsverbandes Vogtland:
http://www.entsorgungsverband-vogtland.de

Environmental Services Association:
http://www.esauk.org/

Hellenic Solid Waste Management Association:
http://www.eedsa.gr/

Le Syndicat Mixte de Traitement et d'Elimination des Déchets:
http://www.dechets79.org

SYCTOM:
http://www.syctom-paris.fr

Ver- und Entsorgungsverband Adelebsen:
http://www.vev-adelebsen.de

Verband Österreichischer Entsorgungsbetriebe:
http://www.voeb.at/

International Associations (selection)

Air & Waste Management Association:
http://www.awma.org

Association of Cities and Regions for Recycling and Sustainable Resource management:
http://www.acrplus.org

FEAD European Federation of Waste Management and Environmental Services:
http://www.fead.be/

International Solid Waste Association:
http://www.iswa.org

Companies/ Industry (selection)

http://www.avfallsverige.se

http://www.eurowaste.net/

http://www.indaver.com

http://www.jakob-becker.de

http://www.preseco.eu

http://www.ragnsells.se

http://www.senternovem.nl/Waste_Management_Department/index.asp

http://www.strautmann-umwelt.de

www.agr.de

www.lobbe.de

Universities (selection)

Cranfield University, Centre for Resource Management and Efficiency:
http://www.cranfield.ac.uk/sas/resource/

Fachhochschule Südwestfalen, IFEU-Institut für Entsorgung und Umwelttechnik:
http://ifeu.typo3-power.de/

HAW Hamburg, CC4E – Competence Center für Erneuerbare Energien und Energieeffizienz:
http://www.haw-hamburg.de/energie.html

RWTH Aachen University, Waste Management Service Centre:
http://www.rwth-aachen.de/go/id/vjs

Portals (selection)

Arbeitsgemeinschaft PVC und Umwelt e.V.:
http://www.agpu.de/

Bundesverband der Deutschen Entsorgungswirtschaft e.V.:
http://www.bde-berlin.org/

Bundesverband Sekundärrohstoffe und Entsorgung e.V.:
http://www.bvse.de/

Bureau International Recycling:
http://www.bir.org/

Deutsche Vereinigung für Wasserwirtschaft, Abwasser und Abfall e.V. (DWA):
http://www.atv.de/portale/dwa_master/dwa_master.nsf/home?readform

European Bioplastics e.V. – Links:
http://www.european-bioplastics.org/index.php?id=192

Hessische Forschungsverbund für Abfall, Umwelt und Ressourcenschutz e.V.:
http://www.uni-kassel.de/hfva/

International Lead and Zink Study Group:
http://www.ilzsg.org/static/usefullinks.aspx?from=1

International Lead Zinc Research Organization Inc.:
http://www.ilzro.org/

Municipal Waste Europe:
http://www.municipalwasteeurope.eu/

Recycling and Waste Management:
http://www.euwid-recycling.com/

Recycling Magazin:
http://www.recyclingmagazin.de/

Verband der Aluminiumrecycling-Industrie e.V. – Partner:
http://www.aluminium-recycling.com/de/verband/partner.php/

Verband Deutscher Metallhändler e.V.:
http://www.metallhandel-online.com/

Waste Management & Research:
http://wmr.sagepub.com/

Waste Management Info Site:
http://www.thinkgreen.com/

Waste Management World:
http://www.waste-management-world.com/

Waste Online:
http://www.wasteonline.org.uk/

2.5 Biofuels: References

2.5.1 References

AmberWaves Nov. 2007 – *Amber Waves* is published four times per year (March, June, September, and December) by the U.S. Department of Agriculture, Economic ResearchService.:
http://www.ers.usda.gov/AmberWaves/November07/Features/Biofuels.htm, last visit: 10 June 2009

Biodiesel 2009, National Biodiesel Board:
http://www.biodiesel.org/pdf_files/fuelfactsheets/prod_quality.pdf last visit: 10 June 2009

Biofuels-platform 2009:
http://www.biofuels-platform.ch
http://www.biofuels-platform.ch/en/infos/bioethanol.php, last visit: 10 June 2009

Biomassenergycentre 2009,The BIOMASS Energy Centre (BEC) is owned and managed by the UK Forestry Commission, via Forest Research, its research agency.
http://www.biomassenergycentre.org.uk/portal/page?_pageid=76,15049&_dad =portal&_schema=PORTAL, last visit: 10 June 2009

DBFZ 2008, Workshop: Biofuels and bio-based chemicals Trieste 18-20 September 2008. – "Second and Third Generation of Biofuels and Biorefineries –Considerations and concepts":
www.ics.trieste.it/VideoStore/BioWorkshop/2008.09.18_16.00-16.59/presenta-tions/7_Thraen.ppt last visit: 10 June 2009

EU 2003, Directive 2003/30/EC of the European Parliament and of the Council of 8 May 2003:
http://ec.europa.eu/energy/res/legislation/doc/biofuels/en_final.pdf

(EurObserv 2008) EurObserv 2008 – The State of Renewable Energies in Europe 8th EurObserv'ER Report, last visit: 10 June 2009

IEA Bioenergy 2008, 2008 IEA Bioenergy: Task 39 'Commercializing First and Second Generation Liquid Biofuels from Biomass:
http://www.task39.org/About/Definitions/tabid/1761/language/en-US/Default. aspx last visit: 10 June 2009

MMG 445 Basic Biotechnology eJournal, "The use of syngas derived from biomass and waste products to produce ethanol and hydrogen" by Joshua D. Mackaluso, MMG 445 Basic Biotechnology eJournal 2007 3: 98-103 www.msu.edu/course/mmg/445/ , last visit: 10 June 2009

MMG 445 Basic Biotechnology eJournal, A review of the processes of biodiesel production- Michael Sheedlo MMG 445 Basic Biotechnology e-Journal 2008 4:61-65
http://ejournal.vudat.msu.edu, last visit: 10 June 2009

OECD/IEA 2008, Agency IEA Bioenergy and Jack Saddler, Warr en Mabee© OECD/IEA, November 2008 From First- to Second-GenerationBiofuel Technologies: An overview of current industry and RD&D activities Ralph Sims, Michael Taylor International Energy

RenewableenergyNor 2009:
http://www.renewableenergy.no/sitepageview.aspx?articleID=177#anker_, last visit: 10 June 2009

SCA BIONORR AB 2009, http://bionorr.se/dokument/Bionorr_eng.pdf, last visit: 10 June 2009

SenterNovem 2009, SenterNovem is an agency under the Ministry of Economic Affairs. Bioethanol in Europe Overview and comparison of production processes Rapport 2GAVE0601:
http://www.senternovem.nl/mmfiles/ECNGAVEbioethanoleindrapport_tcm24-280156.pdf, last visit: 10 June 2009

The Wharton School of the University of Pennsylvania 2007. "Economic and Business Challenges for Biodiesel Production in Turkey". Paul R. Kleindorfer, http://opim.wharton.upenn.edu/risk/library/2007_PRK-UGO_BiodieselTurkey.pdf, last visit: 10 June 2009

Videncenter 2004,
http://www.videncenter.dk/Groenne%20trae%20haefte/Groen_Engelsk/Kap_07.pdf

2.5.2 Tables

Classification of Second Generation Biofuels (OECD/IEA 2008)

2.5.3 Figures

Biodiesel Production Process (Biodiesel 2009)

Conversion routes for sugar or starch feedstocks to ethanol and co-products (OECD/IEA 2008)

Determination of storage tank size (Videncenter 2004)

Ethanol production from lingo-cellulose vie the bio-chemical route (OECD/IEA 2008)

Global biofuel production between 2000 and 2007 (IEA/FO Licht)

Major sugars and sugar polymers for bioethanol production (SenterNovem)

Production of Bioethanol (Biofuels-platform)

Scheme for bioethanol production from starchy raw materials (SenterNovem)

Scheme of a combined sugar/bioethanol production process from sugar beet (SenterNovem)

Thyborøn district heating system (Videncenter 2004)

Wood pellets production process (SCA BIONORR AB)

World ethanol production from first generation and second generation (ligno-cellulose) (Mabee & Saddler 2007)

Worldwide Biofuels Production (FO Licht)

About the Authors

Georgi Chobankov, Bachelor degree at the HAW – Hamburg University of Applied Sciences in "Environmental Engineering". Part time employee in an environmental management consulting company. Research assistant for the RENET Project. 2010 Master degree course in "Renewable Energy Systems – Environmental and Process Engineering" at the HAW – Hamburg University of Applied Sciences.

Faculty of Life Sciences,
Hamburg University of Applied Sciences,
Lohbruegger Kirchstraße 65
21033 Hamburg, Germany,
phone: +49.40.428756267
email: georgiatanasov.chobankov@haw-hamburg.de

M.Sc. Dipl.-Ing. (FH) *Konstantin Haker*. Master of Science in Environmental Engineering. Consultant for the waste management industry. Referee on the yearly VDI-Seminar (VDI = Association of German Engineers) on the Recycling of shredder residues. HAW Project-Manager of the RENET Project. PhD-Studentship at the University of the West of Scotland (UWS) in cooperation with HAW Hamburg regarding Refuse Derived Fuel (RDF) from Shredder Residues.

Faculty of Life Sciences,
Hamburg University of Applied Sciences,
Lohbruegger Kirchstraße 65
21033 Hamburg, Germany,
phone: +49.40.428756267
email: konstantin.haker@haw-hamburg.de

Prof. Dr.-Ing. *Kerstin Kuchta*, Professor at HAW Hamburg University of Applied Sciences since 2002. Professorship for Environmental Management. Research Fields and Lectures: Environmental Management, Environmental Law, Waste and Waste Water Treatment, Ecological Evaluation of processes and products. Memberships and functions: Chairwoman of HAW Solar e.V. (association), board member "Neue Energie Hamburg" e.V., chairwoman of the Research Commission of the HAW, HAW Coordinator of the RENET project, Coordinator of the VDI-Seminar (VDI = Association of German Engineers) on the recycling of shredder residues, Member of the board for VDI – Directive 2343 (Recycling of electronical devices), Dean of the Faculty of Engineering at the German-Kazakh University in Almaty, Kazakhstan.

Faculty of Life Sciences,
Hamburg University of Applied Sciences,
Lohbruegger Kirchstraße 65, Room N5.14,
21033 Hamburg, Germany,
phone: +49.40.428756267
email: kerstin.kuchta@haw-hamburg.de

Professor *Walter Leal Filho* (BS, PhD, DSc, DL) heads the Research and Transfer Centre "Applications of Life Sciences" of the Hamburg University of Applied Sciences, where he is in charge of various sustainability-related projects across the world.

Research and Transfer Centre Applications of Life Sciences,
Faculty of Life Sciences,
Hamburg University of Applied Sciences,
Lohbruegger Kirchstraße 65, Sector S4 / Room 0.38,
21033 Hamburg, Germany,
phone: +49-40-42875-6313, fax: +49-40-42875-6079,
email: walter.leal@.haw-hamburg.de

Franziska Mannke (MBA, BSc., B.A. int.) has been coordinating EU projects on a national and an international level since 2005, with RENET being one of her current projects. In the Research and Transfer Centre "Application of Life Sciences", Franziska is working with the topics Climate Change, Renewable Energies, Innovation and Sustainability as well as development economics and food science related topics.

Research and Transfer Centre Applications of Life Sciences,
Faculty of Life Sciences,
Hamburg University of Applied Sciences,
Lohbruegger Kirchstraße 65,
21033 Hamburg, Germany,
phone: +49-40-42875-6324, fax: +49-40-42875-6079,
email: franziska.mannke@.haw-hamburg.de

Anna Spitsyna got her B.Sc. degree in Industrial ecology in 2000 at Kazakh National Technical University, Kazakhstan, Almaty and then got another degree in "in Sociology and Translation", 2004 (Institute of Language and Translations, Kazakhstan, Almaty). She obtained her Master Degree in Industrial Ecology in 2006 at the Royal Institute of Technology (KTH), Sweden, Stockholm. At the present time she is working in PhD in the areas of "Sustainable Technology" at the Royal Institute of Technology (KTH) and involved in "Renewable Energy Networks Between Turkish and European Universities" (RENET) EU funded project as a project manager at The Royal Institute of Technology (KTH), Sweden, Stockholm.

Royal Institute of Technology (KTH)
Department of Industrial Ecology
Teknikringen 34
114 28 Stockholm
phone: (+46) 08-73 71 50 633
email: annaspi@kth.se
http://www.ima.kth.se/eng/index.htm

Ronald Wennersten is a professor and Head of the Department of Industrial Ecology at Royal Institute of Technology (KTH) in Stockholm, Sweden. He received his Engineering degree and PhD in Chemical Engineering. Ronald Wennersten is working with several EU projects in the area of sustainable development. He developed a close cooperation with several Chinese universities with focus on Industrial Ecology and Sustainable Development and several projects with countries of the former Soviet Union. From 2008 he is coordinating a new European research network for Climate Change Mitigation. Ronald Wennersten has published more than 100 books and articles in international journals and conferences.

Royal Institute of Technology (KTH)
Department of Industrial Ecology
Teknikringen 34
114 28 Stockholm
phone: (+46) 08-790 63 47, (+46) 08-755 66 44
email: rw@kth.se
http://www.ima.kth.se/eng/index.htm

Thematic Index

biodiesel 83–111
bioethanol 83–111

crystalline 113–45

emissions 147–77
energy 147–77

feeding techniques 63–81
fermenters 63–81
filters 147–77
first generation 83–111
fuel 147–77

incineration 147–77

limits 113–45

management 147–77
methanisation 63–81

phases 63–81
plant 113–45

recovery 147–77
renewable energy policy
– Germany 29–62
– laws aiming to increase the production of green energy 29–62
– new restrictions 29–62

– Sweden 29–62
– Turkey 29–62
renewable energy sources in Turkey
– alternatives for reducing the dependence on fossil fuels 13–27
– biomass 13–27
– characteristics 13–27
– geothermal 13–27
– hydropower 13–27
– potential 13–27
– solar 13–27
– wind 13–27

second generation 83–111
separation 147–77
solid biofuels 83–111
substrates 63–81

thermal treatment 147–77
thin film layer 113–45
Turkey
– energy-importing country 13–27
– limitation of domestic fossil fuel resources 13–27

wafer 113–45
waste 2 energy 147–77
waste management 147–77
wastes 147–77

Danyel Reiche (ed.)

Handbook of Renewable Energies in the European Union

Case studies of the EU-15 States
Second, completely revised and updated edition
Forewords by Hermann Scheer, Claude Turmes, and Stephan Kohler
In Collaboration with Mischa Bechberger, Ruth Brand, Matthias Corbach, and Stefan Körner

Frankfurt am Main, Berlin, Bern, Bruxelles, New York, Oxford, Wien, 2005.
330 pp., num. fig. and tab., 1 DVD
ISBN 978-3-631-53560-8 · pb. € 40,10*

This publication is the completely revised and updated second edition of the *Handbook of Renewable Energies*. The handbook is a collection of systematic case studies describing national renewable energy policies in the EU-15.
In all case studies of this edition data from 2003 was used. All the recent developments in the field of renewable energies were integrated, such as new support schemes, for example in Austria, the Netherlands and Sweden, and changing administrative responsibilities as in Germany. As in the first edition, all chapters follow the same structure. At the beginning of each case study, a definition of renewable energies is given for the individual country and the starting position in energy policy as well as the main actors are described. The instruments for promoting renewable energies are shown and each section concludes with an analysis of current obstacles and conditions for future success. Finally, a service chapter informs the reader about the most important associations, websites, and journals pertinent to the subject matter and provides some general information about the EU-15 States.

Contents: Renewable energies in comparison – an analysis by Danyel Reiche and Mischa Bechberger · Renewable energies in the European Union · 15 chapters by different authors outlining the situation of renewable energies in the EU-15 States · and many more

Frankfurt am Main · Berlin · Bern · Bruxelles · New York · Oxford · Wien
Distribution: Verlag Peter Lang AG
Moosstr. 1, CH-2542 Pieterlen
Telefax 0041 (0)32/376 17 27

*The €-price includes German tax rate
Prices are subject to change without notice
Homepage http://www.peterlang.de

www.ingramcontent.com/pod-product-compliance
Ingram Content Group UK Ltd.
Pitfield, Milton Keynes, MK11 3LW, UK
UKHW051855140426
5217IPUK00006B/132